POWER SERIES OVER COMMUTATIVE RINGS

PURE AND APPLIED MATHEMATICS

A Program of Monographs, Textbooks, and Lecture Notes

Contributions to *Lecture Notes in Pure and Applied Mathematics* are reproduced by direct photography of the author's typewritten manuscript. Potential authors are advised to submit preliminary manuscripts for review purposes. After acceptance, the author is responsible for preparing the final manuscript in camera-ready form, suitable for direct reproduction. Marcel Dekker, Inc. will furnish instructions to authors and special typing paper. Sample pages are reviewed and returned with our suggestions to assure quality control and the most attractive rendering of your manuscript. The publisher will also be happy to supervise and assist in all stages of the preparation of your camera-ready manuscript.

LECTURE NOTES
IN PURE AND APPLIED MATHEMATICS

1. *N. Jacobson,* Exceptional Lie Algebras
2. *L.-Å. Lindahl and F. Poulsen,* Thin Sets in Harmonic Analysis
3. *I. Satake,* Classification Theory of Semi-Simple Algebraic Groups
4. *F. Hirzebruch, W. D. Newmann, and S. S. Koh,* Differentiable Manifolds and Quadratic Forms
5. *I. Chavel,* Riemannian Symmetric Spaces of Rank One
6. *R. B. Burckel,* Characterization of C(X) Among Its Subalgebras
7. *B. R. McDonald, A. R. Magid, and K. C. Smith,* Ring Theory: Proceedings of the Oklahoma Conference
8. *Y.-T. Siu,* Techniques of Extension of Analytic Objects
9. *S. R. Caradus, W. E. Pfaffenberger, and B. Yood,* Calkin Algebras and Algebras of Operators on Banach Spaces
10. *E. O. Roxin, P.-T. Liu, and R. L. Sternberg,* Differential Games and Control Theory
11. *M. Orzech and C. Small,* The Brauer Group of Commutative Rings
12. *S. Thomeier,* Topology and Its Applications
13. *J. M. Lopez and K. A. Ross,* Sidon Sets
14. *W. W. Comfort and S. Negrepontis,* Continuous Pseudometrics
15. *K. McKennon and J. M. Robertson,* Locally Convex Spaces
16. *M. Carmeli and S. Malin,* Representations of the Rotation and Lorentz Groups: An Introduction
17. *G. B. Seligman,* Rational Methods in Lie Algebras
18. *D. G. de Figueiredo,* Functional Analysis: Proceedings of the Brazilian Mathematical Society Symposium
19. *L. Cesari, R. Kannan, and J. D. Schuur,* Nonlinear Functional Analysis and Differential Equations: Proceedings of the Michigan State University Conference
20. *J. J. Schäffer,* Geometry of Spheres in Normed Spaces
21. *K. Yano and M. Kon,* Anti-Invariant Submanifolds
22. *W. V. Vasconcelos,* The Rings of Dimension Two
23. *R. E. Chandler,* Hausdorff Compactifications
24. *S. P. Franklin and B. V. S. Thomas,* Topology: Proceedings of the Memphis State University Conference
25. *S. K. Jain,* Ring Theory: Proceedings of the Ohio University Conference
26. *B. R. McDonald and R. A. Morris,* Ring Theory II: Proceedings of the Second Oklahoma Conference
27. *R. B. Mura and A. Rhemtulla,* Orderable Groups
28. *J. R. Graef,* Stability of Dynamical Systems: Theory and Applications
29. *H.-C. Wang,* Homogeneous Banach Algebras
30. *E. O. Roxin, P.-T. Liu, and R. L. Sternberg,* Differential Games and Control Theory II
31. *R. D. Porter,* Introduction to Fibre Bundles
32. *M. Altman,* Contractors and Contractor Directions Theory and Applications
33. *J. S. Golan,* Decomposition and Dimension in Module Categories
34. *G. Fairweather,* Finite Element Galerkin Methods for Differential Equations
35. *J. D. Sally,* Numbers of Generators of Ideals in Local Rings
36. *S. S. Miller,* Complex Analysis: Proceedings of the S.U.N.Y. Brockport Conference
37. *R. Gordon,* Representation Theory of Algebras: Proceedings of the Philadelphia Conference
38. *M. Goto and F. D. Grosshans,* Semisimple Lie Algebras
39. *A. I. Arruda, N. C. A. da Costa, and R. Chuaqui,* Mathematical Logic: Proceedings of the First Brazilian Conference
40. *F. Van Oystaeyen,* Ring Theory: Proceedings of the 1977 Antwerp Conference

41. *F. Van Oystaeyen and A. Verschoren,* Reflectors and Localization: Application to Sheaf Theory
42. *M. Satyanarayana,* Positively Ordered Semigroups
43. *D. L. Russell,* Mathematics of Finite-Dimensional Control Systems
44. *P.-T. Liu and E. Roxin,* Differential Games and Control Theory III: Proceedings of the Third Kingston Conference, Part A
45. *A. Geramita and J. Seberry,* Orthogonal Designs: Quadratic Forms and Hadamard Matrices
46. *J. Cigler, V. Losert, and P. Michor,* Banach Modules and Functors on Categories of Banach Spaces
47. *P.-T. Liu and J. G. Sutinen,* Control Theory in Mathematical Economics: Proceedings of the Third Kingston Conference, Part B
48. *C. Byrnes,* Partial Differential Equations and Geometry
49. *G. Klambauer,* Problems and Propositions in Analysis
50. *J. Knopfmacher,* Analytic Arithmetic of Algebraic Function Fields
51. *F. Van Oystaeyen,* Ring Theory: Proceedings of the 1978 Antwerp Conference
52. *B. Kedem,* Binary Time Series
53. *J. Barros-Neto and R. A. Artino,* Hypoelliptic Boundary-Value Problems
54. *R. L. Sternberg, A. J. Kalinowski, and J. S. Papadakis,* Nonlinear Partial Differential Equations in Engineering and Applied Science
55. *B. R. McDonald,* Ring Theory and Algebra III: Proceedings of The Third Oklahoma Conference
56. *J. S. Golan,* Structure Sheaves over a Noncommutative Ring
57. *T. V. Narayana, J. G. Williams, and R. M. Mathsen,* Combinatorics, Representation Theory and Statistical Methods in Groups: YOUNG DAY Proceedings
58. *T. A. Burton,* Modeling and Differential Equations in Biology
59. *K. H. Kim and F. W. Roush,* Introduction to Mathematical Consensus Theory
60. *J. Banas and K. Goebel,* Measures of Noncompactness in Banach Spaces
61. *O. A. Nielsen,* Direct Integral Theory
62. *J. E. Smith, G. O. Kenny, and R. N. Ball,* Ordered Groups: Proceedings of the Boise State Conference
63. *J. Cronin,* Mathematics of Cell Electrophysiology
64. *J. W. Brewer,* Power Series Over Commutative Rings
65. *P. K. Kamthan and M. Gupta,* Sequence Spaces and Series
66. *T. G. McLaughlin,* Regressive Sets and the Theory of Isols (in press)
67. *H. W. Stech, S. M. Rankin, and T. L. Herdman,* Integral and Functional Differential Equations (in press)
68. *R. Draper,* Commutative Algebra: Analytic Methods (in press)
69. *W. G. McKay and J. Patera,* Tables of Dimensions, Indices, and Branching Rules for Representations of Simple Lie Algebras

Other Volumes in Preparation

POWER SERIES OVER COMMUTATIVE RINGS

James W. Brewer

Department of Mathematics
The University of Kansas
Lawrence, Kansas

MARCEL DEKKER, INC.   New York and Basel

Library of Congress Cataloging in Publication Data

Brewer, James W., 1942-
Power series over commutative rings.

(Lecture notes in pure and applied mathematics ; 64)
Bibliography: p.
Includes index.
1. Commutative rings. 2. Power series rings.
I. Title. II. Series.
QA251.3.B73 512'.4 81-843
ISBN 0-8247-6952-X AACR2

MARCEL DEKKER, INC.

270 Madison Avenue, New York, New York 10016

Current printing (last digit):

10 9 8 7 6 5 4 3 2 1

PRINTED IN THE UNITED STATES OF AMERICA

PREFACE

Formal power series rings have been an object of study for many years, but during the past decade the amount of research done on these rings has taken a quantum leap forward. There are several reasons for this, but regarding the questions treated in this work, the main reason is the impetus provided by R. Gilmer. Several of Gilmer's students have been concerned with the following general problem:

Relate the ideal structure of a ring to that of its power series ring.

Here, we are taking the point of view that for the types of problems under consideration, the answers are known for the polynomial ring. Thus, we seek analogs in the power series case for results known to hold in the polynomial ring case.

We have, for the most part, restricted our attention to power series in a single indeterminate. The reasons are multifarious, but primarily we are concerned with readability and breadth of audience. It is our feeling that by focusing on the single variable case, more people's attention might be gained as well as retained. Consequently, this work is not an encyclopedia into the area of formal power series rings. We have tried to select those results which seem to us to be interesting and appealing as well as admitting of a fairly accessible proof. Sometimes we have left as exercises the easy extension to finitely many variables of results proved for a single variable. Moreover, in Chapter 3, where treating the problem in several variables is the natural approach, we have done exactly that. On the other hand, nowhere have we touched upon power series rings involving infinitely many indeterminates. (The reader desiring information about these rings should consult [16], wherein they are treated quite extensively; at least in so far as their definitions, and there are three distinct ones, and their elementary properties.)

The book is divided into three chapters. Roughly speaking, Chapter 1 contains positive results, Chapter 2 negative results, and Chapter 3 is given over entirely to the study of R-homomorphisms of power series rings. The common feature of the results of Chapter 1 is that at least for Noetherian rings, almost all of the results for polynomial rings will carry over to the power series ring. By contrast, Chapter 2 graphically illustrates how different the situation can be as regards such elementary properties as Krull dimension, transcendence degree, valuation theory, and integrality. Chapter 3 is our favorite because the mathematics it treats is so clean, elegant, and aesthetically pleasing.

We assume a level of sophistication roughly equivalent to that required for Kaplansky's *Commutative Rings* [23], and a knowledge of [23] or of a similar text would be useful, indeed. It must be emphasized, however, that such a knowledge is not essential and that the text can be read by anyone who has had a basic graduate course in algebra. We have intentionally strived for such a level of accessibility, but we are also aware of the obvious danger of boring the expert.

All our rings are commutative and contain an identity element. Wherever possible, we have tried to keep to the terminology and notation of [23]. There are exceptions, but we are confident that our terminology will pose no problems for the reader.

Also following [23], wherein the style is attributed to Landau, each of our results carries the designation "Theorem." The efficiency of this style is evident and more than offsets its defects.

We would like to call attention to the excellent survey article [17] of Gilmer. That paper served as both inspiration and guide throughout much of the writing of the present work. Although published in 1975, [17] was written in 1972, and a great deal of work has been done in the area since 1972. This is particularly true of the questions concerning homomorphisms of power series rings, the subject treated in Chapter 3. In any event, it would not be too far off the mark to say that the present work is nothing more than an expanded updated version of [17].

Much of the material in the text was presented in a seminar at Stellenbosch University in the Republic of South Africa. The audience consisted of Theo Jooste, Louis Le Riche, Eric Martens, Arnold Muller, Mike Sheridan, and Wil van Rooyen. Each of them was a faithful, conscientious, and helpful participant, and I sincerely thank them for their time and comments. I also gratefully acknowledge two monetary grants from the University of Stellenbosch which subsidized trips to Pretoria, where I was able to use the libraries of UNISA and the CSIR.

I would also like to thank the University of Virginia and the University of Kansas. The semester we spent in Charlottesville was profitable in many ways, one of which was the completion of this work. The University of Kansas kindly provided the funds necessary to defray the expenses of final typing.

Finally, I must thank Mrs. Ada Basson, Mrs. Fran Strauss, and Mrs. Judy LaFollette for typing the manuscript and Mrs. Marguerite Ellis for completing the camera copy. They were ever patient with me and for that especially, I am truly grateful.

James W. Brewer

CONTENTS

POWER SERIES OVER COMMUTATIVE RINGS

Chapter 1

BASIC PROPERTIES OF POWER SERIES RINGS

In this chapter we shall establish many of the elementary properties of power series rings. In the process we shall encounter some of the problems with which we shall be concerned in subsequent chapters.

1. Some Basic Facts

The relationship of the power series ring R[[X]] to the coefficient ring R is a special instance of the type of relationship we often encounter in the study of commutative ring theory. Namely, we have two rings S and T and a homomorphism from S to T. The basic problem is to relate the ideal structures of S and T. This is, of course, an intentionally vague problem, but we hope that the reader will come to understand how we are viewing it. In our particular set-up we will seek to relate the ideal theory of R[[X]] to that of R.

We begin with a basic result.

THEOREM 1. The power series $f = \Sigma_{i=0}^{\infty} a_i X^i$ is a unit in R[[X]] if and only if a_0 is a unit in R.

Proof. We produce an inverse $g = \Sigma_{i=0}^{\infty} b_i X^i$ by induction. Define $b_0 = a_0^{-1}$ and having defined $b_0, b_1, \ldots, b_n$ appropriately, we proceed to b_{n+1}. If $fg = 1$, then we must have $0 = \Sigma_{i+j=n+1} a_i b_j$--that is, $a_0 b_{n+1} = -a_1 b_n - \ldots - a_{n+1} b_0$. Thus, we define b_{n+1} to be equal to $a_0^{-1}(-a_1 b_n - \ldots -a_{n+1} b_0)$, which is meaningful since a_0 is a unit of R and $b_0, b_1, \ldots, b_n$ have been previously defined.

At this early juncture we already encounter problems. We disposed of the units of R[[X]] in short order, but the zero-divisors and nilpotent elements are another matter. In contrast to the situation which exists in the polynomial ring R[X], there does not in general exist a characterization of the zero divisors and nilpotent elements of a power series ring. For particular classes of rings a characterization does exist, but that will have to await a subsequent section.

We shall henceforth adopt the following notation. If I is an ideal of the ring R, then I + (X) will denote the ideal of R[[X]] consisting of all those power series whose constant term belong to I. Thus,

$$I + (X) = \{f = \sum_{i=0}^{\infty} a_i X^i \mid a_0 \in I\}$$

One useful way to view I + (X) is as the inverse image of I under the canonical homomorphism from R[[X]] to R which maps a power series in R[[X]] to its constant term.

Armed with this notation and Theorem 1, we easily deduce our next fact.

THEOREM 2. If $\{M_\alpha\}$ is the set of maximal ideals of R, then $\{M_\alpha + (X)\}$ is the set of maximal ideals of R[[X]].

Proof. Let ϕ denote the homomorphism from R[[X]] onto R mapping a power series onto its constant term. If M is a maximal ideal of R, then $M + (X) = \phi^{-1}(M)$ is a maximal ideal of R[[X]]. On the other if $\mathcal{M}$ is a maximal ideal of R[[X]], then the set $M = \{f(0) | f \in \mathcal{M}\}$ is an ideal of R. Here, we are using f(0) to denote the constant term of the power series f. Now, $M \neq R$ for otherwise $\mathcal{M}$ would contain a unit by Theorem 1. Moreover, if $a \in R \setminus M$, then $(\mathcal{M}, a) = R[[X]]$ and so there exist $f \in \mathcal{M}$, $g \in R[[X]]$ such that $1 = g \cdot a + f$. Thus, $1 = g(0) \cdot a + f(0) \in (M,a)$ and it follows that M is a maximal ideal of R. Since M + (X) clearly contains $\mathcal{M}$, the proof is complete.

The above theorem certainly gives a satisfactory description of the maximal ideals of R[[X]] in terms of those of R. Obviously, a much better situation exists here than that which exists between the maximal ideals of R and those of R[X]. What can be said at this early stage about the prime ideals of R[[X]]? That is our next goal. More notation is required.

If I is an ideal of the ring R, then

$$I[[X]] = \{f = \sum_{i=0}^{\infty} a_i X^i | a_i \in I \text{ for all } i\}$$

Then I[[X]] is an ideal of R[[X]] and a useful way to view it is as follows. Let σ denote the canonical homomorphism from R onto R/I. Then σ gives rise to a natural mapping τ from R[[X]] onto (R/I)[[X]] defined by

$$\tau \sum_{i=0}^{\infty} a_i X^i = \sum_{i=0}^{\infty} \sigma(a_i) X^i.$$

It is easy to check that τ is a ring homomorphism and that the kernel of τ is I[[X]].

Warning. I[[X]] must not be confused with IR[[X]], the ideal of R[[X]] generated by the set I. We shall say a good deal more about this later, but it is another instance where things are different from the polynomial ring case.

DEFINITION. If $f = \sum_{i=0}^{\infty} a_i X^i$ is a non zero power series in R[[X]], by the *order* of f we shall mean that nonnegative integer n such that $a_i = 0$ for $i < n$ and $a_n \neq 0$. We shall write $\phi(f)$ for the order of the power series f and if f has order n, we shall call a_n the *initial coefficient* of f.

Certain observations are obvious from the definition.

(1) $\phi(f+g) \geq \min\{\phi(f),\phi(g)\}$ for all f, $g \in R[[X]]$ with equality assured if $\phi(f) \neq \phi(g)$.

(2) $\phi(f\cdot g) \geq \phi(f) + \phi(g)$ for all f, $g \in R[[X]]$.

(3) Suppose that $f = \Sigma_{i=m}^{\infty} a_i X^i$ and $g = \Sigma_{i=n}^{\infty} b_i X^i$ with $a_m \neq 0 \neq b_n$. If a_m or b_n is not a zero-divisor in R, then $\phi(fg) = \phi(f) + \phi(g)$.

As an immediate consequence of (3), we have

THEOREM 3. If D is an integral domain, then D[[X]] is an integral domain.

Our next three results are concerned with prime ideals of R[[X]].

THEOREM 4. Let P be a prime ideal of R. Then P + (X) and P[[X]] are prime ideals of R[[X]].

Proof. The first assertion is clear from previous remarks. As for the second, we have that

$$R[[X]]/P[[X]] \simeq (R/P)[[X]],$$

which is an integral domain by Theorem 3.

THEOREM 5. Suppose that M is a maximal ideal of the ring R. If Q is a prime ideal of R[[X]] containing MR[[X]], then $Q \subseteq M[[X]]$ or Q = M + (X). Therefore, if $M[[X]] \neq \sqrt{MR[[X]]}$, then there exists a prime ideal Q of R[[X]] such that $MR[[X]] \subseteq Q \subsetneq M[[X]]$.

Proof. The set $Q' = \{f(0) \mid f \in Q\}$ is a proper ideal of R, and since $Q \supseteq MR[[X]]$, $Q' \supseteq M$. Hence, $Q' = M$. Suppose that $Q \nsubseteq M[[X]]$ and let $f \in \Sigma_{i=0}^{\infty} a_i X^i \in Q\backslash M[[X]]$. If t is minimal among those positive integers j such that $a_j \notin M$, then $g = \Sigma_{i=0}^{t-1} a_i X^i \in MR[[X]] \subseteq Q$ and therefore $f-g \in Q$. Since $a_0 \in Q' = M$, $t \neq 0$. Thus, $f-g = X^t(\Sigma_{i=t}^{\infty} a_i X^{i-t}) \in Q$ and $\Sigma_{i=t}^{\infty} a_i X^{i-t} \notin Q$ since $a_t \in M$. It follows that $X \in Q$ and that $Q \supseteq M + (X)$, which is maximal.

The second assertion needs no proof.

The following theorem of Kaplansky provides us with an easy method of proving that the Noetherian property is preserved under passage from a ring R to the power series ring R[[X]].

THEOREM 6. Let Q be a prime ideal of R[[X]] and let $Q' = \{f(0) | f \in Q\}$ be the corresponding ideal of R. Then Q is finitely generated if and only if Q' is finitely generated. If Q' is generated by r elements, then Q can be generated by r + 1 elements if $X \in Q$. If $X \notin Q$, then Q can be generated by r elements.

Proof. The ideal Q' is the image of Q under the homomorphism mapping a power series to its constant term. Thus, if Q is finitely generated, so is Q'. Suppose that $Q' = (a_1,\ldots,a_r)$. We distinguish two cases. If $X \in Q$, then $Q = (a_1,\ldots,a_r,X)$. Assume that $X \notin Q$ and let $f_1,\ldots,f_r$ be series with constant terms $a_1,\ldots,a_r$. We claim that $Q = (f_1,\ldots,f_r)$. For take any $g \in Q$. If g has constant term b, then $b = \Sigma_{i=1}^{r} b_i a_i$ for some $b_i \in R$ and $g - \Sigma_{i=1}^{r} b_i f_i = Xg_1$. Here, $Xg_1 \in Q$ and therefore $g_1 \in Q$. In the same way we write $g_1 = \Sigma_{i=1}^{i} c_i f_i + Xg_2$ with $g_2 \in Q$. Continuation of the process leads us to $h_1,\ldots,h_r \in R[[X]]$,

$$h_i = b_i + c_i X + \ldots$$

satisfying $g = h_1 f_1 + \ldots + h_r f_r$.

Recall that a ring is Noetherian if all of its ideals are finitely generated. By a celebrated theorem of Cohen, to prove that a ring is Noetherian, it is sufficient to prove that all the prime ideals are finitely generated. Thus, from Theorem 6, we have the following theorem.

THEOREM 7. If R is Noetherian, so is R[[X]].

2. Zero-divisors

In this section we survey the known results on zero-divisors in the power series ring R[[X]].

We recall a theorem of McCoy which is decisive in the polynomial ring R[X]. A polynomial $f \in R[X]$ is a zero-divisor if and only if there exists a nonzero element $a \in R$ such that $af = 0$ [6].

Unfortunately, there does not, in general, exist an analogue to this theorem for the power series ring. It is instructive to see why, and so we offer the following example.

Example 1. We would like to find a ring R over which we can build a zero-divisor power series $f = \Sigma_{i=0}^{\infty} a_i X^i$ with some a_i a unit of R. If this is possible, we can clearly arrange for a_i to be 1. Let us proceed naively. Obviously, f cannot be a constant. But f can be taken to be a linear polynomial. We have only to find a ring R containing elements a_0 and $\{b_i\}_0^{\infty}$ such that if $g = \Sigma_{i=0}^{\infty} b_i X^i$, then $(a_0+X)\cdot g = 0$. This imposes the following conditions on a_0 and the b's: $a_0 b_0 = 0$ and $b_i + a_0 b_{i+1} = 0$ for $i \geq 0$. Thus, let K be a field with Y and $\{Z_i\}_0^{\infty}$ indeterminates.

Let A be the ideal of $K[Y,\{Z_i\}_0^{\infty}]$ generated by YZ_0 and the elements $\{Z_i + YZ_{i+1}\}_0^{\infty}$. If we set $R = K[Y,\{Z_i\}_0^{\infty}]/A$, $a_0 = Y + A$, and $b_i = Z_i + A$ for $i \geq 0$, it is clear that $(a_0+X)(\Sigma_{i=0}^{\infty} b_i X^i) = 0$ in R[[X]].

Thus, a monic linear polynomial in R[X] can be a zero-divisor in R[[X]]. Clearly, such a polynomial cannot be annihilated by a constant. As we shall soon see, the difficulty lies in the fact that R is not Noetherian.

The analogue of McCoy's Theorem in the Noetherian case is due to D. Fields [13]. We require a technical result which is of independent interest. Recall that an ideal Q of a ring R is said to *primary* for the prime ideal P of R if the following conditions are satisfied: $P = \sqrt{Q}$ and whenever $a \cdot b \in Q$ with $a \notin P$, it follows that $b \in Q$. For other equivalent formulations, see [33].

THEOREM 8. Let R be a ring with Q an ideal of R primary for the prime ideal P. If Q contains a power of P, the Q[[X]] is primary for P[[X]].

Proof. P[[X]] is a prime ideal of R[[X]] and clearly contains Q[[X]]. Suppose that f, $g \in R[[X]]$ with $f \cdot g \in Q[[X]]$. Assuming that $f \notin P[[X]]$, we would like to see that $g \in Q[[X]]$. By working modulo Q[[X]], we may suppose that Q = (0) and then our task is to show that f is not a zero-divisor. Write $f = a_0 + a_1X + \ldots + a_mX^m + \ldots$ where $a_i \in P$ for $0 \leq i \leq m - 1$ and $a_m \notin P$. Since (0) is P-primary, a_i is nilpotent for $0 \leq i \leq m - 1$ and so $h = \Sigma_{i=0}^{m-1} a_iX^i$ is nilpotent. Thus, f is not a zero-divisor if and only if f - h is not a zero-divisor [Ex. 5]. But a_m, the initial coefficient of f - h, is not a zero-divisor of R since $a_m \notin P$. Consequently, f - h is not a zero-divisor.

To complete the proof, we have to show that $P[[X]] \subseteq \sqrt{Q[[X]]}$. This follows from the fact that if k is a positive integer such that $P^k \subseteq Q$, then $(P[[X]])^k \subseteq Q[[X]]$.

If I is an ideal of the ring R, we noted previously that I[[X]] need not be equal to IR[[X]]. It is easy to see that equality does hold if I is finitely generated. In particular I[[X]] = IR[[X]] for all ideals I of R if R is Noetherian. We shall utilize this observation in the proof of the next theorem.

THEOREM 9. Suppose that R is a Noetherian ring. Then there exist nonzero elements $r_1, r_2, \ldots, r_m \in R$ such that f is a zero-divisor of R[[X]] if and only if $r_jf = 0$ for some j.

Proof. Let $(0) = Q_1 \cap \ldots \cap Q_m$ be an irredundant representation of (0) as an intersection of primary ideals of R. If Q_i is P_i-primary, then the set of zero-divisors of R is $\cup_{j=1}^m P_j$ [33] and moreover, there exist nonzero elements $r_1,\ldots,r_m \in R$ such that $r_jP_j = (0)$ for $1 \leq j \leq m$ [23]. In R[[X]], $(0) = Q_1[[X]] \cap \ldots \cap Q_m[[X]]$ and $Q_j[[X]]$ is $P_j[[X]]$-primary by Theorem 8. Clearly, the representation is irredundant and so $\cup_{j=1}^m P_j[[X]]$ is the set of zero-divisors of R[[X]]. Thus, if f is a zero-divisor of R[[X]], $f \in P_j[[X]]$ for some j and $r_jf = 0$.

REMARK. Those readers who feel as if the introduction of primary decomposition is unnecessary should give a proof which does not use it. Such a proof should only use the fact that the set of zero-divisors of R is a finite union of prime ideals, each of which is the annihilator of a single element.

Since the theorem of McCoy holds trivially in case R is an integral domain, it is natural to try to prove it in case R has no nonzero nilpotent elements. We conclude this section with such a result.

THEOREM 10. Suppose that the ring R contains no nonzero nilpotent elements. If f, g $\in$ R[[X]] with fg = 0, then each coefficient of g annihilates f.

Proof. Let c(f) denote the ideal of R generated by the coefficients of f. If c(g) denotes the corresponding ideal for g, our assertion is precisely that $c(f) \cdot c(g) = 0$. Let P be a prime ideal of R. Then, since P[[X]] is prime, either $f \in P[[X]]$ or $g \in P[[X]]$. In particular, if a is an arbitrary coefficient of f and b is an arbitrary coefficient of g, then $a \cdot b \in P$. Thus, $c(f) \cdot c(g) \subseteq P$ for each prime ideal P of R. Since the set of nilpotent elements of R is the intersection of all prime ideals of R, $c(f) \cdot c(g) = (0)$.

REMARK. A ring having no nonzero nilpotent elements is called *reduced*. In this language, a more precise statement of Theorem 10 is the following. If R is a reduced ring and if f is a zero-divisor in R[[X]], then the annihilator of f in R[[X]] is A[[X]], where A is the annihilator in R of the ideal of R generated by the coefficients of f.

3. Nilpotent Elements

In this section we shall investigate the nilpotent elements of R[[X]]. We shall find the situation to be similar to that of the zero-divisor case in that no definitive theorem exists, only strong results in special circumstances. Our first proposition points the way. As is the case with most of the results in this section, it is drawn from Field's paper [13].

THEOREM 11. Let $f = \Sigma_{i=0}^{\infty} a_i X^i \in R[[X]]$. If f is nilpotent, then a_i is nilpotent for $i \geq 0$.

Proof. If P is a prime ideal of R, then P[[X]] is a prime ideal of R[[X]]. Therefore, since f is nilpotent $f \in P[[X]]$ and so $a_i \in P$ for $i \geq 0$. Since P was arbitrary, a_i is nilpotent.

Obviously, the best possible result would be that a power series is nilpotent if and only if each of its coefficients is nilpotent--that is, if the converse of Theorem 11 held. Unfortunately, the converse need not hold as the following example of Fields shows.

Example 2. Let Q be the field of rational numbers, $\{Y_i\}_0^{\infty}$ be a collection of indeterminates, and n be a positive integer, $n \geq 2$. Let $S = Q[\{Y_i\}_0^{\infty}]$ and let I be the ideal of S generated by the set $\{Y_i^n\}_0^{\infty}$. Put $R = S/I$, $a_i = Y_i + I$ for $i \geq 0$ and consider the power series $f = \Sigma_{i=0}^{\infty} a_i X^i \in R[[X]]$. It is clear that $a_i^n = 0$ for each i.

Suppose that f is nilpotent, say $f^m = 0$. Then $a_0^m = 0$ and so $m \geq n$. Let t_1 be the smallest integer k for which a_0^n does not occur in every summand used in computing the kth coefficient of f^m. Since $f^m = 0$, the t_1^{st} coefficient of f^m is 0 and so we have that

$$0 = ca_0^{n-1} a_1^{m-(n-1)} + \text{(other terms, each of which has } a_0^n \text{ as a factor)}$$

and where c is a positive integer. Thus, each of the "other terms" is zero and so $0 = ca_0^{n-1} a_1^{m-(n-1)}$. By the way that R was defined, it follows that $m - (n-1) \geq n$.

We iterate the process as follows. Let t_2 be the smallest integer k for which some summand used in computing the kth coefficient of f^m has neither a_0^n nor a_1^n as a factor. Then we have that

$$0 = ba_0^{n-1} a_1^{n-1} a_2^{m-2(n-1)} + \text{(other terms, each having } a_0^n \text{ or } a_1^n \text{ as a factor)}$$

and where b is a positive integer. As above, it follows that $0 = ba_0^{n-1} a_1^{n-1} a_2^{m-2(n-1)}$ and hence that $m - 2(n-1) \geq n$. By induction, we can argue that for each positive integer k, $m - k(n-1) \geq n$. That is, $m \geq n + k(n-1)$. Since $n \geq 2$, this is clearly absurd and so f cannot be nilpotent.

It is noteworthy that in the above example there is a bound on the orders of nilpotence of the coefficients, but still the power series is not nilpotent.

For which rings does the converse to Theorem 11 hold? As the next result shows, the fact that the ring had characteristic zero was an essential feature of the preceeding example.

THEOREM 12 (Fields). Let R be a commutative ring with characteristic $n = p_1^{e_1} p_2^{e_2} \cdots p_t^{e_t} > 0$, where p_j is a prime integer and $e_j \geq 1$ for $1 \leq j \leq t$. If $f = \Sigma_{i=0}^{\infty} a_i X^i$, then f is nilpotent if and only if there exists a positive integer m such that $a_i^m = 0$ for each $i \geq 0$.

Proof. We first treat the case when n = p is itself a prime integer. Notice that in this case, for any positive integer s, $f^{p^s} = \Sigma_{i=0}^{\infty} a_i^{p^s} X^{i p^s}$. Thus, if f is nilpotent of order h and if we choose s such that $p^s > h$, then $0 = a_i^{p^s}$ for $i \geq 0$. The argument can be easily reversed.

In the general set-up, for $1 \leq j \leq t$, let ϕ_j be the natural homomorphism from R[[X]] to $R[[X]]/p_jR[[X]]$. (Here, p_jR is the ideal of R generated by the element p_j1 of R. In particular, the ring R/p_jR has characteristic p_j).

If f is nilpotent, then for $1 \leq j \leq t$, $\phi_j(f)$ is nilpotent in $(R/p_jR)[[X]]$ and by the prime characteristic case, there is a positive integer m_j such that $0 = [\phi_j(a_i)]^{m_j} = \phi_j(a_i^{m_j})$ for all i. Set $m = m_1e_1 + \ldots + m_te_t$. Then for $i \geq 0$, we have that

$$a_i^m = (a_i^{m_1})^{e_1} \ldots (a_i^{m_t})^{e_t} \in (p_1R)^{e_1} \ldots (p_tR)^{e_t} = (0)$$

Conversely, if such an m exists, then for $1 \leq j \leq t$, $\phi_j(a_i^m) = [\phi_j(a_i)]^m = 0$ for $i \geq 0$. By the prime characteristic case, there is a positive integer m_j such that $[\phi_j(f)]^{m_j} = 0$. Again setting $m = m_1e_1 + \ldots + m_te_t$, we have that

$$f^m = (f^{m_1})^{e_1} \ldots (f^{m_t})^{e_t} \in (p_1R[[X]])^{e_1} \ldots (p_tR[[X]])^{e_t} = (0)$$

Although the assumption concerning the characteristic of the ring in Theorem 12 might be surprising, that we can also prove a similar result when R is Noetherian should not be. The proof is quite elementary. We take this opportunity to formally introduce some terminology and notation.

If R is a ring and if $f \in R[[X]]$, then we denote by c(f) the ideal of R generated by the coefficients of f. We shall refer to c(f) as the *content* of the power series f.

THEOREM 13. Let R be a Noetherian ring with $f = \Sigma_{i=0}^{\infty} a_iX^i$ an element of R[[X]]. Then f is nilpotent if and only if there is an integer m such that $a_i^m = 0$ for $i \geq 0$.

Proof. Since R is Noetherian, c(f) is finitely generated. Therefore, if f is nilpotent, each a_i is nilpotent by Theorem 11 and since c(f) is finitely generated c(f) is also nilpotent. If $[c(f)]^m = (0)$, then $a_i^m = 0$ for all i. On the other hand, if $a_i^m = 0$ for $i \geq 0$, then there is a positive integer k such that $[c(f)]^k = (0)$. It is easy to see that $f^k = 0$.

The astute reader will no doubt observe that the crucial point in the above argument is that c(f) is finitely generated. Thus, if R is any ring and if $f \in R[[X]]$ is such that c(f) is finitely generated, then f is nilpotent if and only if the ideal c(f) is nilpotent.

We conclude this section with a lemma of J. Arnold [1]. Some appreciation for the result can be gained by studying the example at the beginning of the section.

THEOREM 14. Let I be an ideal of the ring R and suppose that there is a positive integer k such that $x^k = 0$ for each $x \in I$. Then there is a positive integer m such that $mI^m = (0)$. Moreover, each element of I[[X]] is nilpotent.

Proof. Suppose that we have integers $r, s_1, \ldots, s_t$ such that $ra_1^{s_1} \ldots a_t^{s_t} = 0$ for all $a_1, \ldots, a_t \in I$. (Certainly, this holds for $r = t = 1$ and $s_1 = k$). Suppose that some s_i is greater than one and we may assume then that $s_1 \geq 2$. Then if $b_0, b_1, \ldots, b_t \in I$, we have that

$$0 = r(b_0+b_1)^{s_1} b_2^{s_2} \ldots b_t^{s_t} = rb_0^{s_1-2}(b_0+b_1)^{s_1} b_2^{s_2} \ldots b_t^{s_t} = \sum_{j=0}^{s_1} \alpha_j$$

where $\alpha_j = r\binom{s_1}{j} b_0^{2s_1-j-2} b_1^j b_2^{s_2} \ldots b_t^{s_t}$. If $0 \leq j \leq s_1-2$, then $2s_1 - j - 2 \geq s_1$ so that $\alpha_j = 0$. Also, $\alpha_{s_1} = b_0^{s_1-2}(rb_1^{s_1} \ldots b_t^{s_t}) = 0$.

From these considerations we see that

$$0 = \alpha_{s_1-1} = rs_1 b_0^{s_1-1} b_1^{s_1-1} b_2^{s_2} \ldots b_t^{s_t}.$$

Thus, by adding an element, we can reduce two exponents. Therefore, by a finite number of repetitions of this procedure, we can reduce all exponents to 1 and we will produce positive integers m_1 and m_2 so that $m_1 a_1 a_2 \ldots a_{m_2} = 0$ for all elements $a_1, a_2, \ldots, a_{m_2} \in I$. If we set $m = m_1 m_2$, then $mI^m = (0)$. Now, let $f \in I[[X]]$, $f = \Sigma_{i=0}^{\infty} a_i X^i$.

If the m chosen above can be taken to be a prime p, then $f^{p^k} = \Sigma_{i=0}^{\infty} a_i^{p^k} X^{ip^k} = 0$. If m is not prime, say $m = p_1^{e_1} \cdots p_t^{e_t}$, p_i prime, then let $\phi_j : R[[X]] \to (R/_{p_j} I^{p_j})[[X]]$ be the canonical homomorphism for $1 \leq j \leq t$. By the prime case, we have that $0 = [\phi_j(f)]^{p_j^k}$. That is, $(f^{p_j})^k \in p_j I^{p_j}[[X]]$. If we put $n = [(p_1^k)e_1 + \cdots + (p_t^k)e_t]m$, then we have that

$$\begin{aligned} f^n &= [((f)^{p_1^k})^{e_1} \cdots ((f)^{p_t^k})^{e_t}]^m \\ &\in [(p_1 I^{p_1})^{e_1} [[X]] \cdots (p_t I^{p_t})^{e_t} [[X]]]^m \\ &= mI^m [[X]] = (0) \end{aligned}$$

REMARK. The reader will note that the first assertion of Theorem 14 is really a statement about nil ideals with a bound on the order of nilpotence of their elements.

4. Properties Stable Under Passage From D to D[[X]]

In this section we shall devote most of our attention to power series rings over an integral domain. Our goal is to indicate which classical properties of integral domains are preserved under passage to the power series ring and which properties are not. Note that Theorem 3 assures that this is a meaningful task.

We begin with the simplest case of all.

THEOREM 15. Let K be a field. Then $\{(X^n)\}_{n=1}^{\infty}$ is the set of all nonzero ideals of K[[X]]. Thus, K[[X]] is a principal ideal domain having a unique maximal ideal--that is, a discrete (rank one) valuation domain.

Proof. Let I be a nonzero ideal of K[[X]] and choose $f \in I$ to have order minimal among all nonzero elements of I. If $f = \Sigma_{i=m}^{\infty} a_i X^i$ with $a_m \neq 0$, then $f = X^m \Sigma_{i=0}^{\infty} a_{i+m} X^i$ and $\Sigma_{i=0}^{\infty} a_{i+m} X^i$ is a unit in K[[X]] by Theorem 1. Thus, $X^m \in (f) \subseteq I \subseteq (X^m)$. The second assertion is obvious for (X) is clearly the unique maximal ideal of K[[X]].

We weren't really expecting the property of being a field to pass to the power series ring. However, the discrete valuation domain (DVR) K[[X]] is so fundamental to our study that it could not be relegated to the exercises. Our next results will not be so trivial. We recall the relevant definitions.

DEFINITION. Let D be an integral domain with quotient field K and let t be an element of K. The element t is said to be *integral* over D if t is a root of a monic polynomial having coefficients in D. The element t is said to be *almost integral* over D if there exists an element $d \in D$, $d \neq 0$, such that $dt^n \in D$ for all positive integers n. The integral domain D is said to be *integrally closed* if each element of K integral over D belongs to D. D is said to be *completely integrally closed* if each element of K almost integral over D belongs to D.

It is easy to prove that a completely integrally closed domain is integrally closed and any valuation domain of rank greater than one provides an example of an integrally closed domain which is not completely integrally closed [16]. Now each of these two properties is stable under polynomial ring formation, and the following theorems delineate precisely the situation which exists with regard to power series ring formation. In this form we have lifted them directly from Bourbaki [7] via Gilmer [16].

THEOREM 16. If D is an integral domain, then D[[X]] is completely integrally closed if and only if D is completely integrally closed.

Proof. Let K be the quotient field of D. Then $D = D[[X]] \cap K$ and since the intersection of completely integrally closed domains is completely integrally closed, one of the assertions is clear.

Let L be the quotient field of D[[X]]. Now K[[X]], being a factorial ring is completely integrally closed. Therefore, to prove that D[[X]] is completely integrally closed whenever D is, we must show that if $f = \Sigma_{i=0}^{\infty} a_i X^i \in K[[X]] \cap L$ and if, as an element of L, f is almost integral over D[[X]], then $f \in D[[X]]$. Thus, there is a nonzero element $g = \Sigma_{i=m}^{\infty} b_i X^i$ where $b_m \neq 0$ having the property that $gf^n \in D[[X]]$ for all positive integers n. We prove by induction that $a_i \in D$ for each i. Assuming that $a_0, \ldots, a_{j-1}$ are in D, where j is a nonnegative integer, then

$$f^* = a_0 + \ldots + a_{j-1} X^{j-1} \in D[[X]]$$

and

$$g(f-f^*)^n \in D[[X]] \text{ for each positive integer } n$$

If $a_j = 0$, $a_j \in D$. If $a_j \neq 0$, then for each positive integer n,

$$g(f-f^*)^n = \left(\sum_{i=m}^{\infty} b_i X^i\right) \left(\sum_{i=j}^{\infty} a_i X^i\right)^n$$

$$= b_m a_j^n X^{m+nj} + \text{(terms of higher order)}$$

Hence, $b_m a_j^n \in D$ for each n and so a_j is almost integral over D. Since D is completely integrally closed, $a_j \in D$.

THEOREM 17. Suppose that D is an integral domain containing a nonunit t such that $\bigcap_{n=1}^{\infty} t^n D \neq (0)$. Then D[[X]] is not integrally closed.

Proof. Let $b = t^{-1}$. By induction, it is straightforward to prove the existence of a sequence $\{n_i\}_0^\infty$ of integers such that the power series $f = \Sigma_{i=0}^\infty n_i b^{2i+1} X^i$ satisfies the equation $f(t-Xf) = 1$. In particular, $n_0 = 1$ so that $f \in K[[X]] \setminus D[[X]]$, where K denotes the quotient field of D. Also, f belongs to the quotient field of D[[X]], for if d is a nonzero element of $\cap_{n=1}^\infty t^n D$, then

$$df = \sum_{i=0}^{\infty} n_i \ (db^{2i+1}) \ X^i \in D[[X]]$$

Finally, $(Xf)^2 - t(Xf) + X = 0$, so that Xf is an element of the quotient field of D[[X]], not in D[[X]], which is integral over D[[X]].

It follows from Theorem 17 that if V is a valuation domain of rank two, then V is integrally closed but V[[X]] is not. To exhibit an appropriate t one need only chose any nonunit of V which does not lie in the height-one prime ideal of V.

So, we see that there is a class of integrally closed domains which is well-behaved under power series ring formation, but that the class of all integrally closed domains itself is not. For those classes of integrally closed domains of classical ideal theory which are stable under polynomial ring formation, such as factorial rings and Krull domains, it is meaningful to consider the power series extension.

P. Samuel [30] in a famous example, proved that a power series ring over a factorial ring need not be factorial. A nice treatment of this topic appears in Fossum [15]. The situation is better for Krull domains. We recall the definition.

DEFINITION. Let D be an integral domain with quotient field K. D is said to be a *Krull* domain if there exists a family $\{V_\alpha\}$ of valuation domains having quotient field K and satisfying the following properties:

(1) $D = \bigcap_\alpha V_\alpha$

(2) Each V_α is a discrete (rank one) valuation domain

(3) For each $t \in K$, $t \neq 0$, t is a unit in all but a finite number of the domains V_α

(4) For each α, there exists a prime ideal P_α of D such that $V_\alpha = D_{P_\alpha}$

We call the set $\{V_\alpha\}$ a *defining family* for the Krull domain D. There is, of course, a considerable body of literature devoted to the study of Krull domains. The reader who desires to refresh his memory on the subject should consult Bourbaki [7] or Gilmer [16]. A more encyclopedic reference is Fossum's book [15].

THEOREM 18. If D is a Krull domain, so is D[[X]].

Proof. Again our proof is drawn from Bourbaki [7] via Gilmer [16]. We first note that if V is a discrete rank one valuation domain, then $(V[[X]])[1/X] = (V[[X]])_{\{X^i\}_1^\infty} = V'$ is a Euclidean domain. To prove this, let v be a valuation associated with V; we assume that the value group of v is Z, the additive group of integers. Each nonzero element $f \in V'$ can be uniquely expressed in the form $X^k(a_0+a_1X+\dots)$ for some integer k, where $a_0 \neq 0$. We define $\phi: V'\setminus\{0\} \to Z$ by $\phi(f) = v(a_0)$. We claim that ϕ is a Euclidean algorithm on V', for if $f = X^k(a_0+a_1X+\dots)$ and $g = X^r(b_0+b_1X+\dots)$ are nonzero elements of V', then

$$\phi(fg) = v(a_0b_0) = v(a_0) + v(b_0) \geq v(a_0) = \phi(f)$$

To produce elements $s, t \in V'$ such that $f = sg + t$, where $t = 0$ or $\phi(t) < \phi(g)$, we can assume that $\phi(f) \geq \phi(g)$. Thus, $a_0/b_0 \in V$ and

$f - (a_0/b_0)X^{k-r}\,g$ is either zero or is of the form $X^{r_1}(c_0+c_1X+\ldots)$, where $r_1 > k$ and $c_0 \neq 0$. If $f - (a_0/b_0)X^{k-r}\,g = 0$ or if, in the second case, $v(c_0) < v(b_0)$, then we can take $s = (a_0/b_0)X^{k-r}$ and $t = f - (a_0/b_0)X^{k-r}\,g$. If the second occurs and $v(c_0) \geq v(b_0)$, we repeat the process, subtracting $(c_0/b_0)X^{k_1-r}\,g$ from $f - (a_0/b_0)X^{k-r}\,g$, obtaining

$$f - [(a_0/b_0)X^{k-r} + (c_0/b_0)X^{k_1-r}]g$$

By induction we conclude that either there is an integer i for which we can take

$$s = (a_0/b_0)X^{k-r} + (c_0/b_0)X^{k_1-r} + \ldots + (d_0/b_0)X^{k_i-r}$$

and $t = f - sg$ or g divides f in V' and we can take $s = f/g$, $t = 0$. This justifies the claim.

To prove the theorem itself, let $\{V_\alpha\}$ be a defining family for D and let K be the quotient field. It is straightforward to verify that

$$D[[X]] = K[[X]] \cap (\cap_\alpha V_\alpha[[X]][1/X])$$

K[[X]] is a rank one discrete valuation domain by Theorem 15 and hence is trivially a Krull domain. We have just observed that $V_\alpha[[X]][1/X]$ is Euclidean and therefore also a Krull domain. Moreover, if $f = \Sigma_{i=m}^{\infty} a_iX^i$ is a nonzero element of D[[X]], with $a_m \neq 0$, then a_m is a nonunit in only finitely many of the V_α's and hence f is a nonunit of only finitely many of the domains $V_\alpha[[X]][1/X]$. It is now an exercise (cf Exercise 7] to show that D[[X]] is indeed a Krull domain.

Exercises

1. State and prove the appropriate analogues of Theorems 1, 2, 3 and 4 for the power series ring $R[[X_1,\ldots,X_n]]$. What about analogues for Theorems 5 and 6?
2. Prove that if R is Noetherian, so is $R[[X_1,\ldots,X_n]]$.
3. Argue directly that any principal ideal domain is completely integrally closed.
4. Give an example of a GCD-domain D such that D[[X]] is not a GCD-domain. See [23] for the definition.
5. Let R be a ring with x an element of R. *Prove*: If y is a nilpotent element of R, then x is not a zero-divisor if and only if x - y is not a zero-divisor.
6. Prove directly that the ring R of Example 1 is not Noetherian.
7. Let $\{D_\alpha\}$ be a family of integral domains which are subrings of a field L, let $D = \cap_\alpha D_\alpha$ and assume that each nonzero element of D is a nonunit in only finitely many of the D_α's. *Prove*: If each D_α is a Krull domain, then D is a Krull domain.
8. Prove that if D is a principal ideal domain, then D[[X]] is a factorial ring. (Hint: Use a theorem of Kaplansky which says that R is factorial if and only if each nonzero prime ideal of R contains a nonzero principal prime.)

CHAPTER 2

PATHOLOGY IN R[[X]]

In the previous chapter we made reference to Samuel's example of a factorial domain D such that D[[X]] is not factorial. Indeed, we saw in Chapter 1 several examples of pathological behavior in the power series ring R[[X]]. As the title of this chapter indicates, very few results of a positive nature are proved here and rather, we accentuate the negative. We shall investigate the quotient field of D[[X]], valuation domains between D[[X]] and its quotient field, integral dependence between pairs of power series rings, and global dimension of power series rings, and "power invariance".

We begin with what is surely the most interesting instance of pathology treated in this chapter. Namely, we begin by proving that there are rings of finite Krull dimension whose power series rings are infinite dimensional.

5. Dimension Theory of R[[X]]

In this section we shall treat the problem of computing the Krull dimension of R[[X]] given that the Krull dimension of R is known. In order to appreciate the power series situation, it is essential that we understand the polynomial ring case. For that reason, we shall dwell longer than is our custom upon the polynomial ring R[[X]]. Recall the relevant definitions.

DEFINITION. Let R be a ring. If there exists a chain $P_0 \subsetneq P_1 \subsetneq \cdots \subsetneq P_n$ of n + 1 prime ideals of R, but so such chain of n + 2 prime ideals, then R is said to *have dimension n*. In this case we write dim R = n. If there is no bound on the lengths of chains of prime ideals of R, then R is said to be *infinite dimensional*.

It is immediate from the definition that if R is infinite dimensional, so are R[X] and R[[X]]. This follows from the fact that if P is a prime ideal of R, then P[X] and P[[X]] are prime ideals of R[X] and R[[X]] respectively. Consequently, we shall only be concerned with the finite dimensional case.

The fundamental result needed for computing the dimensional of R[X] in terms of that of R is the following [cf 23]. If $P_1 \subsetneq P_2 \subsetneq P_3$ is a chain of three distinct prime ideals of R[X], then $P_1 \cap R \subsetneq P_3 \cap R$.

From this fact it follows at once that if R has dimension n, then $n + 1 \leq \dim (R[X]) \leq 2n + 1$. Moreover, examples show that all intermediate values may be obtained by appropriately choosing R--that is, if m and n are nonnegative integers with $n + 1 \leq m \leq 2n + 1$, then there exists a ring R such that dim R = n and dim (R[X]) = m.

For the power series case no such theorem exists. In fact, as we shall see, there exist zero-dimensional rings whose power series rings have infinite dimension.

The major contribution to this problem is the paper [1] of Arnold and we shall reproduce its content almost in its entirety. The paper is difficult, albeit elementary, and sadly we've been unable to simplify it. Thus, the reader might choose to omit the proof of the main theorem on the first reading. Bearing this in mind, we shall try to convey as much information as possible in the text proper.

Before dealing with the pathology, we present a positive result which acts as a limiting factor on how bad things can get. As might be expected, the Noetherian rings are well-behaved.

THEOREM 19. Let R be a Noetherian ring of dimension n. Then R[[X]] has dimension n + 1.

Proof. If $P_0 \subsetneq P_1 \subsetneq \cdots \subsetneq P_n$ is a chain of n + 1 prime ideals of R, then $P_0[[X]] \subsetneq P_1[[X]] \subsetneq P_2[[X]] \subsetneq \cdots \subsetneq P_n[[X]] \subsetneq P_n + (X)$ is a chain of n + 2 prime ideals of R[[X]]. Therefore, dim R[[X]] $\geq$ n + 1.

Conversely, let M + (X) be an arbitrary maximal ideal of R[[X]], where M is maximal in R. Then the height of M is less than or equal to n and so by [23] there exist $a_1, \ldots, a_n \in M$ such that M is minimal over $(a_1, \ldots, a_n)R$. But then M + (X) is minimal over $(a_1, \ldots, a_n, X)$ in R[[X]] and so by Krull's Principal Ideal Theorem [23], the height of M + (X) is less than or equal to n + 1.

Now for the negative results. We begin with a definition.

DEFINITION. Let I be an ideal of the ring R. We shall call I an SFT-ideal (for "strong finite type") if there exists a finitely generated ideal $B \subseteq I$ and a positive integer k such that $a^k \in B$ for each element $a \in I$. If S is a ring, then we shall call S an A-ring if each ideal of S is an SFT-ideal.

We now state the main theorem.

THEOREM 20. Let R be a ring. The following conditions are equivalent.

(1) There exists an ideal I of R such that $I[[X]] \subsetneq \sqrt{IR[[X]]}$.

(2) There exists a prime ideal P of R such that $P[[X]] \neq \sqrt{PR[[X]]}$.

(3) R is not an A-ring.

Moreover, if one of these conditions is satisfied, then R[[X]] is infinite dimensional.

Proof. (1) $\Rightarrow$ (2): If I is an ideal of R such that $[[X]] \subsetneq \sqrt{IR[[X]]}$, then there exists a prime ideal Q or F[[X]] such that $IR[[X]] \subseteq Q$ but $I[[X]] \not\subseteq Q$. If $P = Q \cap R$, then $P \supseteq I$ and so

$P[[X]] \supseteq I[[X]]$. Therefore, $Q \supseteq PR[[X]]$, but $Q \not\supseteq P[[X]]$. It follows that $P[[X]] \neq \sqrt{PR[[X]]}$.

(2) $\Rightarrow$ (3): Suppose that B is an ideal of R which is an SFT-ideal. By definition, there exists a positive integer k and a finitely generated ideal $C \subseteq B$ such that $b^k \in C$ for each element $b \in B$. Denoting R/C by $\overline{R}$ and B/C by $\overline{B}$, it follows from Theorem 14 that each element of $\overline{B}[[X]]$ is nilpotent in $\overline{R}[[X]]$. Therefore, if $g \in B[[X]]$, then $g \in \sqrt{C[[X]]} = \sqrt{CR[[X]]} \subseteq \sqrt{BR[[X]]}$. Consequently, if P is a prime ideal of R such that $P[[X]] \neq \sqrt{PR[[X]]}$, then P is not an SFT-ideal and R cannot be an A-ring.

(3) $\Rightarrow$ (1): This is the implication whose proof is long and involved.

In the process of proving the implication, we shall also show that the dimension of R[[X]] is infinite. Suppose that (3) holds and let M be an ideal of R which is not an SFT-ideal. Then we may choose a sequence $\{a_i\}_0^\infty$ of elements of M such that $a_{n+1}^{n+1} \notin (a_0, a_1, \ldots, a_n)$ for each nonnegative integer n. Set $I_n = (a_0, \ldots, a_n)$ and set $I = \bigcup_{n=0}^{\infty} I_n$. We now proceed to build an infinite matrix $\{a_{i,j}\}$ as follows. For $i = 1$, take $a_{i,j} = a_j$ for $0 \leq j < \infty$. Having defined the sequence $\{a_{i,j}\}_{j=0}^{\infty}$ for $1 \leq i < n$, define $a_{n,j} = a_{n-1,\, j^2+1}$ for each nonnegative integer j. Thus, $\{a_{i,j}\}$ is a matrix having countably many rows and countably many elements in each row. The exact form of the recursive relation is, for the most part, of little import. In what follows, we most often use the fact that in the matrix, elements occuring early in row $i + 1$ occur late in row i.

We find it convenient at this point to introduce some additional notation. The reader should neither be dismayed nor put off by it, for although cumbersome, it is not a difficult notion with which to work.

Suppose that $g \in R[[X]]$, $g = \sum_{i=0}^{\infty} b_i X^i$, and let n, m, μ, and r be integers such that $m \geq n \geq 1$ and $r \geq 0$. We shall say that the tuple (g, m, μ, r) *has property* (n) if for $i \geq r$, there exists an integer t_i such that the following hold:

(i) $b_{t_i} = a^{\mu}_{m,i} + \alpha$ for some $\alpha \in I_{s_i-1}$,

(ii) $t_i \leq \mu k_i$, and

(iii) $b_j \in I_{s_i-1}$ for $0 \leq j < t_i$.

Here, we are assuming that in the matrix $\{a_{\gamma,\sigma}\}$ previously defined,

$$a_{m,i} = a_{n,k_i} = a_{1,s_i}.$$

We first prove several properties which are formal consequences of the definition.

CLAIM 1. Suppose that n and n_1 are positive integers with $n \geq n_1$. If (g,m,μ,r) has property (n), then (g,m,μ,r) has property (n_1).

Proof. Properties (i) and (iii) of the definition are independent of the choice of n. Thus, we have only to verify that (ii) holds. Suppose that $i \geq r$ and that $a_{m,i} = a_{n,k_i} = a_{n_1,v_i}$. Then $k_i \leq v_i$ and hence $t_i \leq \mu k_i \leq \mu v_i$. Therefore, the same choice of t_i will work to prove that (g,m,μ,r) has property (n_1).

CLAIM 2. Suppose that m and m_1 are positive integers with $m_1 \geq m$. If (g,m,μ,r) has property (n), then (g,m_1,μ,r) has property (n).

Proof. Suppose that $g = \sum_{i=0}^{\infty} b_i X^i$ and let $i \geq r$. If $a_{m_1,i} = a_{m,j_i} = a_{n,k_i} = a_{1,s_i}$, then since $j_i \geq i \geq r$, there exists an integer t_{j_i} such that:

(i) $b_{t_{j_i}} = a^{\mu}_{m,j_i} + \alpha$ for some $\alpha \in I_{s_i-1}$,

(ii) $t_{j_i} \leq \mu k_i$, and

(iii) $b_\lambda \in I_{s_i-1}$ for $0 \leq \lambda < t_{j_i}$.

Taking $\tau_i = t_{j_i}$ and using the fact that $a_{m_1,i} = a_{m,j_1}$, we see that τ_i satisfies the properties (i), (ii) and (iii) of the definition and so (g,m_1,μ,r) has property (n).

Since our eventual goal is to construct a long chain of prime ideals in R[[X]], we are interested in building a long nested sequence of multiplicatively closed sets in R[[X]]. This we now begin to do.

For $n \geq 1$, define $S_n = \{g \in R[[X]] \mid (g,m,\mu,r)$ has property (n) for some choice of m, μ and $r\}$. If for $n \geq 1$, we also define $f_{(n)} = \sum_{j=0}^{\infty} a_{n,j}X^j$, then one sees easily that $(f_{(n)},n,1,0)$ has property (n). To verify this, one can even take $t_i = i$ for $i \geq 0$. Thus, $f_{(n)} \in S_n$ and so $S_n \neq \emptyset$. Moreover, it is immediate from Claim 1 that if n and n_1 are positive integers with $n \geq n_1$, then $S_n \subseteq S_{n_1}$.

A somewhat more difficult result is the following.

CLAIM 3. For $n \geq 1$, S_n is multiplicatively closed.

Proof. Let g and $h \in S_n$, where $g = \sum_{i=0}^{\infty} b_iX^i$ and $h = \sum_{i=0}^{\infty} c_iX^i$. Suppose that (g,m_1,μ_1,r_1) and that (h,m_2,μ_2,r_2) have property (n). By Claim 2, we may assume that $m_1 = m_2$ and clearly, we may assume that $r_1 = r_2$. Thus, we write $m_1 = m_2 = m$ and $r_1 = r_2 = r$. We shall show that $(gh,m,\mu_1+\mu_2,r)$ has property (n). Suppose that $i \geq r$ and that $a_{m,i} = a_{n,k_i} = a_{1,s_i}$. By assumption, there exists integers t_i and τ_i such that $b_{t_i} = a_{m,i}^{\mu_1} + \alpha$ and $c_{\tau_i} = a_{m,i}^{\mu_2} + \beta$ for some $\alpha,\beta \in I_{s_i-1}$. Moreover, b_λ, $c_\delta \in I_{s_i-1}$ for $0 \leq \lambda < t_i$ and $0 \leq \delta < \tau_i$. We shall show that $t_i + \tau_i$ works for gh. If $gh = \sum_{j=0}^{\infty} \xi_jX^j$, then $\xi_{t_i+\tau_i} = b_{t_i}c_{\tau_i} + \sum_{\lambda+\delta=t_i+\tau_i,\ \lambda\neq t_i,\ \delta\neq\tau_i} b_\lambda c_\delta$. But if $\lambda \neq t_i$ and $\delta \neq \tau_i$, then either $\lambda < t_i$ or $\delta < \tau_i$. Consequently, either $b_\lambda \in I_{s_i-1}$ or

$c_\delta \in I_{s_i-1}$. Since $b_{t_i} c_{\tau_i} = a_{m,i}^{\mu_1+\mu_2} + \alpha a_{m,i}^{\mu_2} + \beta a_{m,i}^{\mu_1} + \alpha\beta$, it follows that $\xi_{t_i+\tau_i} = a_{m,i}^{\mu_1+\mu_2} + \gamma \in I_{s_i-1}$. Also, we have that $t_i \leq \mu_1 k_i$ and that $\tau_i \leq \mu_w k_i$. Hence, $t_i + \tau_i \leq (\mu_1+\mu_2)k_i$. Finally, if $0 \leq \lambda < t_i + \tau_i$, then $\xi_\lambda = \Sigma_{j=0}^{\lambda} b_j c_{\lambda-j} \in I_{s_i-1}$ since either $j < t_i$ or $\lambda - j < \tau_i$.

One final technical lemma is needed.

CLAIM 4. If n and n_1 are positive integers with $n > n_1$ and if $g \in S_n$, then $g + h f_{(n_1)} \in S_n$ for arbitrary $h \in R[[X]]$.

Proof. Suppose that $g = \Sigma_{j=0}^{\infty} b_j X^j$ and that (g,m,μ,r) has property (n). Let $\eta = \min\{\text{all nonnegative integers } i \mid a_{m,i} = a_{n,k_i} \Rightarrow k_i \geq \mu\}$. Set $r_1 = \mathrm{Max}\{r,\mu\}$. If $q = g + h f_{(n_1)} = \Sigma_{j=0}^{\infty} \xi_j X^j$, then we wish to show that (q,m,μ,r_1) has property (n). Thus, suppose that $i \geq r_1$ and that $a_{m,i} = a_{n,k_i} = a_{v,\lambda_i} = a_{1,s_i}$. By assumption, there exists an integer t_i such that $b_{t_i} = a_{m,i}^{\mu} + \alpha$ for some $\alpha \in I_{s_i-1}$ and such that $t_i \leq \mu k_i \leq k_i^2$. Since $\lambda_i \geq k_i^2 + 1$ and $t_i \leq k_i^2$, it follows that $\lambda_i > t_i$. Thus, for $0 \leq j \leq t_i < \lambda_i$, we have that $a_{v,j}$ is "to the left" of $a_{v,\lambda_i} = a_{1,s_i}$. Therefore, $a_{v,j} \in I_{s_i-1}$. (Note that here we *have* used the precise form of the recursive relation to say that $\lambda_i \geq k_i^2 + 1$.) Consequently, if $h = \Sigma_{j=0}^{\infty} c_j X^j$ and $h f_{(n_1)} = \Sigma_{j=0}^{\infty} \gamma_j X^j$, then $\gamma_{t_i} = \Sigma_{j=0}^{t_i} a_{v,j} c_{t_i-j} \in I_{s_i-1}$. Therefore, $\xi_{t_i} = t_{t_i} + \gamma_{t_i} = a_{m,i}^{\mu} + \alpha + \gamma_{t_i}$ and (i) of the definition is satisfied. We already have that $t_i \leq \mu k_i$, so that (ii) is also satisfied. To see that (iii) holds, suppose that $0 \leq \delta < t_i$. By assumption, we have that $b_\delta \in I_{s_i-1}$. Also $\gamma_\delta = \Sigma_{j=0}^{\infty} a_{v,j} c_{\delta-j} \in I_{s_i-1}$ since $j \leq \delta < t_i \leq k_-^2$ implies that

$a_{v,j} \in I_{s_i-1}$. Consequently, $\xi_\delta = b_\delta + \gamma_\delta \in I_{s_i-1}$ and the claim is justified.

We come at last to the conceptual part of the proof. We proceed by induction to construct our long chain of prime ideals and begin by showing that $IR[[X]] \cap S_1 = \emptyset$. Thus, let $g \in IR[[X]] \cap S_1$.

$$g = \sum_{j=0}^{\infty} b_j X^j$$

Then there is a finitely generated ideal $C \subseteq I$ of R such that the content of g, $c(g)$, is contained in C. By definition of I, there exists a nonnegative integer k such that $c(g) \subseteq I_k$. Suppose that (g,m,μ,r) has property (1) and that r has been chosen so that if $i \geq r$ and $a_{m,i} = a_{1,s_i}$, then $s_i > \max\{\mu,k\}$. If t_i is such that $b_{t_i} = a_{m,i}^{\mu} + \alpha$ for some $\alpha \in I_{s_i-1}$, then we have that $a_{m,i}^{\mu} + \alpha \in I_k \subseteq I_{s_i-1}$. Therefore, $a_{m,i}^{\mu} \in I_{s_i-1}$, a contradiction, for long ago we chose a_{1,s_i} so that $a_{1,s_i}^{s_i} = a_{m,i}^{s_i} \notin I_{s_i-1}$ and we have that $\mu < s_i$. Thus, $IR[[X]] \cap S_1 = \emptyset$. Note that this also proves (1) for $f_{(1)} \in S_1$ and so $f_{(1)} \in I[[X]] \setminus \sqrt{IR[[X]]}$. Now, $S_1 \cap IR[[X]] = \emptyset$ implies the existance of a prime ideal P_1 of $R[[X]]$ such that $P_1 \supseteq IR[[X]]$ and $P_1 \cap S_1 = \emptyset$ [23]. Suppose that there exists a chain $P_1 \subsetneq P_2 \subsetneq \cdots \subsetneq P_n$ of prime ideals of $R[[X]]$ such that $P_n \cap S_n = \emptyset$ and let $C_n = P_n + (f_{(n)})$. If $g \in S_{n+1}$, then by Claim 4, $g + h f_{(n)} \in S_{n+1} \subseteq S_n$ for arbitrary $h \in R[[X]]$. It follows that $g + h f_{(n)} \notin P_n$ and one sees easily then that $g \notin C_n$. Thus, $C_n \cap S_{n+1} = \emptyset$ and so there exists a prime ideal P_{n+1} of $R[[X]]$ such that $P_{n+1} \supseteq C_n$ and $P_{n+1} \cap S_{n+1} = \emptyset$. Now, $f_{(n)} \in S_n$ and $P_n \cap S_n = \emptyset$. Consequently, $C_n \supsetneq P_n$ and the proof is now complete.

THEOREM 21. Suppose that M is a maximal ideal of a ring R such that M is not an SFT-ideal. Then there exists an infinite chain of prime ideals of R[[X]], $Q_0 \subsetneq Q_1 \subsetneq \ldots$ such that $Q_i \cap R = M$ for each i. Moreover, the height of M[[X]] is infinite.

Proof. The first assertion follows directly from the last paragraph of the proof and the fact that M is maximal. As for the second, since $Q_i \cap R = M$, $Q_i \supseteq MR[[X]]$ for each i. Therefore, by Theorem 5, $Q_i \subseteq M[[X]]$.

As we have repeatedly noted, for an arbitrary ring R it need not be true that IR[[X]] = I[[X]] for each ideal I of R. In fact, this condition holds if and only if R is Noetherian [See Exercise 4]. At this point we prove a much stronger version due to Arnold, Gilmer, and Heinzer [5].

THEOREM 22. Let R be a ring and suppose that for each prime ideal P of R, P[[X]] = PR[[X]]. Then R is Noetherian.

Proof. We first prove that R has Noetherian prime spectrum--i.e. that R satisfies the ascending chain condition on radical ideals. The proof is reminiscent of an argument from the proof of Theorem 20. Suppose that $A_1 \subsetneq A_2 \subsetneq A_3 \subsetneq$ is a strictly ascending chain of radical ideals of R. Set $A = \cup_{i=1}^{\infty} A_i$ and choose for each i an element $a_i \in A \setminus A_i$. If $\alpha = \Sigma_{i=1}^{\infty} a_i X^{i!}$ and if k is a positive integer, then the coefficient of $X^{k(i!)}$ in α^k is a_i^k for $i \geq k$. Since A_i is a radical ideal of R, $a_i^k \notin A_i$. Thus, α is an element of A[[X]], but no power of α belongs to AR[[X]] since $AR[[X]] \subseteq \cup_{i=1}^{\infty} A_i[[X]]$. Consequently, $\alpha \in A[[X]] \setminus \sqrt{AR[[X]]}$ and so there exists a prime ideal Q of R[[X]] such that $AR[[X]] \subseteq Q$ but $A[[X]] \not\subseteq Q$. Let $P = Q \cap R$. Then $PR[[X]] \subseteq Q$ and $A[[X]] \subseteq P[[X]]$. It follows that $PR[[X]] \neq P[[X]]$ and that R has Noetherian prime spectrum.

By a result of J. Ohm and R. Pendleton [24], since R has Noetherian spectrum, each prime ideal of R is the radical of a finitely generated ideal. Thus, if R is not Noetherian, there exists a prime ideal P of R such that P is maximal with respect to not being finitely generated [23] and P is the radical of a finitely generated ideal A of R.

We claim that $P^n \subseteq A$ for some positive integer n. To see this, we will first have to isolate the precise condition required in order that PR[[X]] = P[[X]]. It is easy to verify that

(*) PR[[X]] = P[[X]] if and only if for each countably generated ideal I of R with $I \subseteq P$ there exists a finitely generated ideal B of R with $I \subseteq B \subseteq P$.

To justify the claim, suppose that for each positive integer m, $P^m \not\subseteq A$. Then for each positive integer m there exist elements $q_{m1}, q_{m2}, \ldots, q_{mm} \in P$ such that $q_{m1}, q_{m2}, \ldots, q_{mm} \notin A$. Let Q be the ideal of R generated by the countable set $\bigcup_{m=1}^{\infty}\{q_{m1}, q_{m2}, \ldots, q_{mm}\}$. Then $Q \subseteq P$ and no power of Q is contained in A. But by (*) there exists a finitely generated ideal B of R with $Q \subseteq B \subseteq P$. Since P is the radical of A, there exists a positive integer m such that $B^m \subseteq A$. It follows that $Q^m \subseteq A$, a contradiction.

Next, consider the (R/P)-module P/P^2. Since P has property (*), P/P^2 has the property that given a countably generated submodule E of P/P^2, there exists a finitely generated submodule F of P/P^2 such that $E \subseteq F$. Now the ring R/P is Noetherian since P was chosen to be maximal among non finitely generated ideals of R. If P/P^2 is not finitely generated (R/P)-module then it contains a countably generated submodule E wich is not finitely generated. As remarked above, there exists a finitely generated (R/P)-module F such that $E \subseteq F$. Hence, F is a Noetherian (R/P)-module and E is finitely generated. We conclude that P/P^2 is a finitely generated (R/P)-module.

Thus, there exists a finitely generated ideal C of R with $C \subseteq P$ and $P = P^2 + C$. Then $P^2 = P^3 + PC$ and so $P = P^2 + C = P^3 + PC + C = P^3 + C$. By induction.

$$P = P^m + C \text{ for each positive integer } m.$$

But $P^n \subseteq A$ and so $P = P^n + C \subseteq A + C \subseteq P$. It follows that P is finitely generated and the proof is complete.

Thus, R is Noetherian if and only if P[[X]] = PR[[X]] for each prime ideal P of R. Now by Theorem 19 the dimension of R[[X]] is well-behaved relative to that of R if R is Noetherian and by Theorem 20, it is badly behaved if R fails to be an A-ring. Stated positively, Theorem 20 says that R is an A-ring if and only if $P[[X]] = \sqrt{PR[[X]]}$ for each prime ideal P of R. Therefore, in some sense the pathology occuring in the dimension of R[[X]] can be attributed to the fact that P[[X]] is much "larger" than PR[[X]]. We note that it is still an open problem to determine whether or not the A-rings are precisely the rings R for which dim R = n implies that dim(R[[X]]) = n + 1. We will say a bit more about this later. For now, we return to the problem of constructing examples.

By means of Theorem 20, we can construct many interesting examples of finite dimensional rings whose power series rings are infinite dimensional. Our task is, of course, to find rings which are not A-rings. It is obvious that if a prime ideal P of a ring R is an SFT-ideal, then P is the radical of a finitely generated ideal. Thus, if each prime ideal of R is an SFT-ideal, then each prime ideal of R is the radical of a finitely generated ideal. Such rings are precisely the rings having Noetherian prime spectrum [24]--that is, the rings satisfying the ascending chain condition on radical ideals. For reference sake, we explicitly record this observation.

THEOREM 23. If R is an A-ring, then R has Noetherian prime spectrum.

The converse to Theorem 23 fails. Let V be a rank one nondiscrete valuation domain. This is a valuation domain having only one nonzero prime ideal M with $M = M^2$. (There are many ways to construct such rings and the reader is again directed to [16] for a nice treatment of the topic of valuation domains and their ideal theory). Obviously, M cannot be principal and so M is not an SFT-ideal, but that V has Noetherian prime spectrum is clear. We are now in a position to present the example previously promised. Recall that a ring R is said to be *von Neumann regular* if and only if R is zero-dimensional and reduced. One can prove that a von Neumann regular ring has Noetherian prime spectrum if and only if it is a direct sum of fields. In particular, if R is an infinite direct product of fields, then R[[X]] is infinite dimensional.

Let R be the ring of Example 2 in Chapter 1. Then $M = (a_0, a_1, \ldots)$ is the unique prime ideal of R and, as we proved, $f = \Sigma_{i=0}^{\infty} a_i X^i$ is not nilpotent. If g is an element of MR[[X]], then $g = \Sigma_{i=0}^{t} a_i h_i$ for some positive integer t and $h_i \in R[[X]]$. Since $a_i^n = 0$ for $0 \leq i \leq t$, g is nilpotent. Therefore, we see that $f \in M[[X]] \setminus \sqrt{MR[[X]]}$. By Theorem 20, R[[X]] is infinite dimensional.

Now each of the above examples is a zero-dimensional ring. Note however, that if V is a rank one non-discrete valuation domain, then the dimension of V[[X]] is infinite by an earlier remark. Thus, a one-dimensional integral domain may have an infinite dimensional power series ring.

There are several questions concerning this problem which remain unsettled and we feel as if we should explicitly mention some of them. We conclude this section by doing so.

Although Theorem 20 is a beautiful result, it is unsatisfactory in some respects. Specifically, is a form of the converse true? That is, if R is a finite dimensional A-ring, is R[[X]] finite dimensional? If R is Noetherian, the answer is "yes" by Theorem 19, and so we are really asking how closely A-rings approximate Noetherian rings, at least in so far as their behavior with respect to power series rings. Since an A-ring does have Noetherian prime spectrum, this question is not as far fatched as it might appear at first blush. Perhaps an investigation of A-rings themselves would be in order.

Another unsettled question is the following. If R is an n-dimensional ring and if the dimension of R[[X]] is finite, is the dimension of R[[X]] equal to n + 1? In every case in which the dimension has been computed, the answer is "yes." Fields in [14] calculated that the dimension of V[[X]] was n + 1 whenever V was an n-dimensional (discrete) valuation domain and Arnold extended this considerably in [2] when he showed that if D is an n-dimensional Prüfer A-domain then D[[X]] has dimension n + 1.

A problem obviously related to this latter problem is the problem of determining the dimension of R[[X]] when R is an n-dimensional member of a certain nice class of rings, say the class of all Krull domains or the class of all factorial rings. At present, these problems all seem to be quite difficult.

6. Valuation Overrings of D[[X]]

Let D be an integral domain with P a prime ideal of D. It is easy to prove that if D_P is a valuation domain, then $(D[X])_{P[X]}$ is also a valuation domain. Indeed, a very useful, albeit not well-known fact, is that each valuation overring of D[X] which is a localization of D[X] is either of this form or of the form $(K[X])_{(f)}$, where K is the quotient field of D and f is an irreducible polynomial in K[X]. In this section we shall see that results of this kind do not extend to D[[X]] and we shall find that, as in the previous section, the failure of P[[X]] to be equal to PD[[X]] seems to be the principal obstruction.

The material of this section is drawn from [4] and because it would be difficult to improve the exposition in [4], we have altered it only slightly. Finally, since it will play a key role in the proofs, the reader is once again referred to [16] for a treatment of the requisite valuation theory.

Let V be a rank one valuation domain with M the maximal ideal of V. If K is the quotient field of D and if v is a valuation on K associated with V, then we may take the value group of v to be a subgraoup of the additive group of real numbers. For $f = \Sigma_{i=0}^{\infty} a_iX^i \in V[[X]]$, we define

$$v^* (f) = \inf_{0 \le i < \infty} \{v(a_i)\}$$

THEOREM 24. If v^* is defined as above, then v^* is a valuation on $V[[X]]$. Moreover, $MV[[X]]$ is a prime ideal of $V[[X]]$ and $(V[[X]])_{MV[[X]]}$ is the valuation domain associated with v^*.

Proof. Let $f = \Sigma_{i=0}^{\infty} a_i X^i$ and $g = \Sigma_{i=0}^{\infty} b_i X^i$ belong to $V[[X]]$. It is straightforward to verify that $v^*(f+g) \geq \min\{v^*(f), v^*(g)\}$. To show that $v^*(fg) = v^*(f) + V^*(g)$, it suffices to show that $v^*(fg) \leq v^*(f) + v^*(g)$. We first consider the case in which there exist coefficients a_r and b_s of f and g respectively, with the property that $v^*(f) = v(a_r)$ and $v^*(g) = v(b_s)$. If r and s are minimal with this property then

$$v(\sum_{i+j=r+s} a_i b_j) = v^*(f) + v^*(g)$$

and since $(\Sigma_{i+j=r+s} a_i b_j)$ is the coefficient of X^{r+s} in fg, it follows that $v^*(fg) \leq v^*(f) + v^*(g)$. In particular, if v is a discrete valuation, this shows that v^* is a valuation. Therefore, suppose that v is a nondiscrete valuation, let $f = \Sigma_{i=0}^{\infty} a_i X^i$ and $g = \Sigma_{i=0}^{\infty} b_i X^i$ be arbitrary elements of $V[[X]]$, and choose $m \in M \setminus (0)$. If $f' = \Sigma_{i=0}^{\infty} a_i m^i X^i$, then either f' is a polynomial or $\lim_{n\to\infty} v(a_n m^n) = \infty$. In either case, there exists an integer s such that $v^*(f') = v(a_s m^s)$. Consequently, if we also set $g' = \Sigma_{i=0}^{\infty} b_i m^i X^i$, then by the case previously handled, $v^*(f'g') \leq v^*(f') + v^*(g')$. But,

$$v^*(fg) = \inf_{v(m)>0} \{v^*(f'g')\} \leq \inf_{v(m)>0} \{v^*(f')+v^*(g')\} = v^*(f) + v^*(g)$$

Thus, v^* is a valuation. Since $MV[[X]] = \{f \in V[[X]] | v^*(f) > 0\}$, $MV[[X]]$ is the center of the valuation domain associated with v^* and therefore, is a prime ideal.

We now prove that $(V[[X]])_{MV[[X]]}$ is the valuation domain associated with the valuation v^*. Since V is a rank one valuation domain, $V[[X]]$ is integrally closed by Theorem 16. Thus, $(V[[X]])_{MV[[X]]}$ is integrally closed and so it suffices to prove that v^* is the unique valuation on V[[X]] centered on MV[[X]] [16]. Let w be such a valuation. If v' is the restriction of w to K, then since V is the valuation domain associated with v', we may assume that $v' = v$. If $f \in V[[X]] \setminus MV[[X]]$, then $w(f) = 0 = v^*(f)$. If $f \in MV[[X]]$ and if $v^*(f)$ is an element of the value group of v, then there exists an element $m \in M$ such that $v^*(f) = v(m)$ and hence there exists an element $f' \in V[[X]] \setminus MV[XX]]$ such that $f = mf'$. Consequently, $w(f) = w(m) + w(f') = v(m) = v(f)$. Now suppose that $f \in MV[[X]]$ and that $v^*(f) = \eta$ is not an element of the value group of v. For $\varepsilon > 0$, let $m_\varepsilon \in M$ be such that $v(m_\varepsilon) = \eta_\varepsilon$, where $0 < \eta - \eta_\varepsilon < \varepsilon$. Since $v^*(f) > \eta_\varepsilon$, there exists $f_\varepsilon \in V[[X]]$ such that $f = m_\varepsilon f_\varepsilon$. Thus, $w(f_\varepsilon) + v(m_\varepsilon) > \eta_\varepsilon$ and it follows that $w(f) \geq \eta = v^*(f)$. Now, let α be an element of the value group of v, $\alpha > \eta$, and let $g \in V[[X]]$ be such that $v^*(g) = \alpha - \eta$. Then $v^*(fg) = \alpha$, so by a previous case, $v^*(f) + v^*(g) = v^*(fg) = w(fg) = w(f) + w(g)$. Since $w(f) \geq v^*(f)$ and $w(g) \geq v^*(g)$, it follows that $v^*(f) = w(f)$. Therefore v^* and w agree on each element of V[[X]].

We now state the principal result of this section.

THEOREM 25. Let D be an integral domain with quotient field K and let P be a prime ideal of D. If $(D[[X]])_{P[[X]]}$ is a valuation domain, then D_P is a DVR. Moreover, $(D[[X]])_{P[[X]]}$ is a DVR.

Proof. Since $D_P = (D[[X]]_{P[[X]]} \cap K$, D_P is a valuation domain. If the height of P is greater than one, then the rank of D_P is greater than one. Therefore, there exists a prime ideal Q of D such that $QD_P \subsetneq PD_P$ and such that $QD_P = \cap_{n=1}^{\infty} a^n D_P$ for some $a \in P \setminus Q$. We show show that $(D[[X]])_{Q[[X]]}$ is not a valuation domain by exhibiting an element ξ, belonging to the quotient field of D[[X]], such that neither ξ nor ξ^{-1} belongs to $(D[[X]])_{Q[[X]]}$. Let $q \in Q \setminus (0)$. Then $q \in \cap_{n=1}^{\infty} a^n D_P$, so that for each positive integer i, there exist elements r_i, $s_i \in D$, $s_i \notin P$ such that $q = (r_i/s_i)a^i$. Set $v_n = \Sigma_{j=0}^{n} j$ and let $\xi = \Sigma_{i=0}^{\infty} (s_{v_i}/a^{v_i})X^i$. Then $q\xi = \Sigma_{i=0}^{\infty} (qs_{v_i}/a^{v_i})X^i =$

$\Sigma_{i=0}^{\infty} r_{v_i} X^i \in D[[X]]$ and therefore, $\xi$is an element of the quotient field of $D[[X]]$. Suppose that $\xi = h/g$, where h and g are elements of $D[[X]]$, $h = \Sigma_{i=0}^{\infty} h_i X^i$ and $g = \Sigma_{i=0}^{\infty} g_i X^i$. We show that each of h and g must be elements of $Q[[X]]$. Now $\xi g = {}_{v} h$, so that for each nonnegative integer k we have that $\Sigma_{i+j=k} (g_i s_{v_j})/a^{v_j} = h_k$ and hence that

$$g_0 s_{v_k} = a^{v_k} (h_k - \sum_{i+j=k, i\neq 0} (g_i s_{v_j}/a^{v_j}))$$

$$= a^{v_k} h_k - \sum_{i+j=k, i\neq 0} a^{v_k - v_j} g_i s_{v_j} \in a^k D$$

Thus, $g_0 \in a^k D_p$ for each positive integer k and it follows that $g_0 \in (\cap_{k=1}^{\infty} a^k D_P) \cap D = Q$. Suppose that we have shown that $g_i \in Q$ for $0 \leq i < n$. For an arbitrary positive integer k, consider the equation $\Sigma_{i+j=n+k} (g_i s_{v_j}/a^{v_j}) = h_{n+k}$. We see from this that

$$g_n s_{v_k} = a^{v_k}[h_{n+k} - \sum_{\substack{i+j=n+k \\ i>n}} (g_i s_{v_j}/a^{v_j}) - \sum_{\substack{i+j=n+k \\ i<n}} (g_i s_{v_j}/a^{v_j})]$$

$$= a^{v_k} h_{n+k} - \sum_{\substack{i+j=n+k \\ i>n}} a^{v_k - v_j} g_i s_{v_j} - a^{v_k} \sum_{\substack{i+j=n+k \\ i<n}} (g_i s_{v_j}/a^{v_j})$$

For $0 \leq i < n$, $g_i \in Q$. Therefore, $g_i s_{v_j} \in (a^{v_j + k}) D_P$ and hence $(g_i s_{v_j})/a^{v_j} \in a^k D_P$. It now follows that $g_n s_{v_k} \in a^k D_P$. Thus $g_n \in (\cap_{k=1}^{\infty} a^k D_P) \cap D = Q$. By induction, $g \in Q[[X]]$. To see that $h \in Q[[X]]$, we note that for any nonnegative integer n, $h_n = \Sigma_{i+j=n} (g_i s_{v_j})/a^{v_j}$.

From a repetition of the above argument, we conclude that $h_n \in Q$. It follows that neither ξ nor ξ^{-1} belongs to $(D[[X]])_{Q[[X]]}$. However, since $Q[[X]] \subseteq P[[X]]$ and $(D[[X]])_{Q[[X]]}$ is a valuation domain, this is a contradiction. It follows that P is a height one prime and that D_P is a rank one valuation domain.

Suppose now that D_P is non-discrete and let v be a valuation on K associated with D_P. For notational purposes, put $D_P = V$, $PD_P = M$, and let v* denote the extension to V[[X]] as defined in Theorem 24. Choose $m \in P \setminus (0)$ and $f \in P[[X]] \setminus MV[[X]]$. We show that neither m/f nor f/m belongs to $(D[[X]])_{P[[X]]}$. If m/f = g/h with g, $h \in D[[X]]$, then $mh = fg \in MV[[X]]$, which is a prime ideal by Theorem 24. Therefore, $g \in MV[[X]] \cap D[[X]] \subseteq M[[X]] \cap D[[X]] = P[[X]]$. Now $v^*(g) = v^*(f) + v^*(g) = v^*(mh) = v(m) + v^*(h) \geq v(m)$. Thus, there exists g' V[[X]] such that $g = mg'$ and hence, $h = fg' \in P[[X]]$. We conclude that D_P is discrete. Furthermore, since D_P is discrete, MV[[X]] = M[[X]]. By Theorem 24, $(V[[X]])_{M[[X]]}$ is a DVR and hence $(V[[X]])_{M[[X]]} \cap$ (quotient field of D[[X]] is a DVR centered on P[[X]]. But $(D[[X]])_{P[[X]]}$ is a valuation domain and therefore equals $(V[[X]])_{M[[X]]} \cap$ (quotient field of D[[X]]). This completes the proof.

Thus in order that $(D[[X]])_{P[[X]]}$ be a valuation domain it is necessary that D_P be a DVR. However, the converse need not hold. That is, there exists an integral domain D having a prime ideal P such that D_P is a DVR, but such that $(D[[X]])_{P[[X]]}$ is not a valuation domain.

Example. Recall that an integral domain R is said to be an *almost Dedekind* domain if R_Q is a DVR for each prime ideal Q of R. This notion has proven to be a useful one, particularly in the are of constructing counterexamples. Of course, the non-Noetherian ones are the most interesting. Thus, let D be an almost Dedekind domain which is not Noetherian. (We refer the reader to [16] where such domains are constructed.) Then there exists a maximal ideal M of D such that M is not finitely generated. In fact, since D is one-dimensional and D_M is a DVR, M is not even the radical of a finitely generated ideal and so, in particular, M is not an SFT-ideal. By Theorem 21,

M[[X]] has infinite height, which it could not have by virtue of Theorem 25 if $(D[[X]])_{M[[X]]}$ were to be a valuation domain. Thus, $(D[[X]])_{M[[X]]}$ is not a valuation domain.

A second conclusion which one draws from Theorem 25 is that if D_P is not a DVR, then $(D[[X]])_{P[[X]]}$ can never be a valuation domain. Therefore, the situation is considerably different from the polynomial ring case. However, in case D_P is a DVR, the pathology does, as we stated before, seem to stem from the fact that PD[[X]] need not be equal to P[[X]]. That is a consequence of the final theorem of this section.

THEOREM 26. Suppose that P is a prime ideal of the integral domain D such that D_P is a DVR. If PD[[X]] is a prime ideal of D[[X]], then $(D[[X]])_{PD[[X]]}$ is a DVR.

Proof. Let $PD_P = qD_P$, where $q \in P$ and let $f = \Sigma_{i=1}^{n} t_i h_i$ be an element of PD[[X]], where $t_i \in P$, $h_i \in D[[X]]$. Since $t_i D_P \subseteq qD_P$ for $1 \leq i \leq n$, it follows that there exists $f^{(1)} \in D[[X]]$ and an $s_1 \in D \backslash P$ such that $f = (qf^{(1)})/s_1$. If $f^{(1)} \in PD[[X]]$, then we may repeat the procedure. Moreover, if this process failed to terminate, then for each positive integer n, we would obtain an element $f^{(n)} \in D[[X]]$ and an element $s_n \in D \setminus P$ such that $f = (q^n f^{(n)})/s_n$. Therefore, $f \in \cap_{n=1}^{\infty} a^n (D_P[[X]])_{PD_P[[X]]} \cap$ (the quotient field of D[[X]]) = (0). Thus, if $f \neq 0$, then there exists a positive integer n such that $f = (q^n h)/s$, where $h \in D[[X]] \setminus PD[[X]]$ and $s \in D \setminus P$. To prove that $(D[[X]])_{PD[[X]]}$ is a valuation domain, it suffices to show that given f and g belonging to PD[[X]], then either f/g or g/f belongs to $(D[[X]])_{PD[[X]]}$. But this is clear from the above argument. In fact, we have shown that each element of $(D[[X]])_{PD[[X]]}$ is of the form uq^r, where u is a unit of $(D[[X]])_{PD[[X]]}$ and r is a nonnegative integer. There $(D[[X]])_{PD[[X]]}$ is a DVR.

7. The Quotient Field of D[[X]]

The reader will have noticed that throughout the previous section we used the rather cumbersome notation "quotient field of D[[X]]." The reason for that will become clear in this section for in this section we shall be interested in the quotient fields of power series rings over integral domains. The trivial case offers little resistance.

THEOREM 27. Let K be a field. Then

$$K((X)) = \{\sum_{i=m}^{\infty} a_i X^i \mid a_i \in K, \text{ m an integer}\}$$

$$= (K[[X]])\,[1/X]$$

is the quotient field of K[[X]].

Proof. By Theorem 15, K[[X]] is a DVR with unique maximal ideal (X). In (K[[X]])[1/X] we have inverted X and so (0) is the only surviving prime ideal. The result now follows.

Therefore, if D is an integral domain with quotient field K, we have the following set-up, where L denotes the quotient field of D[[X]].

$$\begin{array}{ccc} K[[X]] & \text{-----} & K((X)) \\ | & & | \\ D[[X]] & \text{-----} & L \end{array}$$

In this section, we shall investigate this situation and find it to be remarkably difficult. This is particularly noticable when contrasted with the polynomial ring case.

Suppose that R is any ring and that S is a multiplicatively closed subset of R. It is well-known and quite useful that $(R[X])_S \simeq R_S[X]$. This fact can be especially useful in case D is an integral domain with, say, quotient field K. For then we have that $(D[X])_{D*} = K[X]$, where D* denotes the set of nonzero elements of D. Thus, we may utilize facts about the Euclidean domain K[X] to deduce facts about D[X]. If the analogous result held for power series, then the set-up given above would trivialize and there would be no problem. This is just a fancy way of saying that although the quotient field of D[X] is K(X), it is not always the case that K((X)) is the quotient field of D[[X]]. In fact, a clever observation of Gilmer [18] shows that it is rarely the case. We state the result in the more general form given in [32], but our proof is that of Gilmer.

THEOREM 28. Let D be an integral domain with S a multiplicatively closed subset of D. The following conditions are equivalent:

(1) the quotient field of $D_S[[X]]$ = the quotient field of D[[X]].

(2) $D_S[[X]] \subseteq (D[[X]])_{D*}$.

(3) For each sequence $\{s_i\}_{i=0}^{\infty}$ of elements of S, $\bigcap_{i=0}^{\infty} s_iD \neq (0)$.

Proof. It is clear that (2) implies (1).

(1) ⇒ (3): Let $\{s_i\}_{i=0}^{\infty}$ be a sequence of elements of S. For each nonnegative integer k, we define $c_k = (s_0s_1 \ldots s_k)^{-1}$. Then $c = \Sigma_{i=0}^{\infty} c_iX^i \in D_S[[X]]$ and so there exists an element $d \in D*$ and a power series $f \in D[[X]]$ such that $dc = f$. Therefore, for each non-negative integer m, $d(s_0s_1 \ldots s_m)^{-1} \in D$. It follows that $d \in s_mD$ and so di is a nonzero element of $\bigcap_{i=0}^{\infty} s_iD$.

(3) ⇒ (2): Let $\xi \in D_S[[X]]$, $\xi = \Sigma_{i=0}^{\infty} \xi_iX^i$. For each integer $m \geq 0$ there exists an element $s_m \in S$ such that $s_m\xi_m \in D$. By (3), there exists an element $s \in \bigcap_{i=0}^{\infty} s_iD$, with $s \neq 0$. Hence $s\xi \in D[[X]]$ and so $\xi \in (D[[X]])_{D*}$.

We also explicity record the motivating case.

THEOREM 29. Let D be an integral domain with quotient field K and let L denote the quotient field of D[[X]]. The following conditions are equivalent:

(1) $L = K((X))$.

(2) $K[[X]] = (D[[X]])_{D^*}$.

(3) $\{a_i\}_{i=0}^{\infty}$ is a sequence of nonzero elements of D, then

$\cap_{i=0}^{\infty} a_i D \neq (0)$.

The conditions of Theorem 29 are sufficiently stringent so as to make one wonder whether or not D[[X]] and K[[X]] can ever share the same quotient field, but, as Gilmer shows in [18], this can happen when D is a valuation domain. The idea is to build a valuation domain whose rank is of uncountable ordinal type which has no height one prime ideal and then condition (3) of Theorem 29 may be checked.

So it is rare that D[[X]] and K[[X]] have the same quotient field. We next ask, "If the quotient field of K[[X]] is larger than that of D[[X]], then how much larger?" This question was essentially settled by P. Sheldon in a nice piece of work published in 1971. More precisely, Sheldon is concerned with the following problem. Suppose that D is an integral domain and a is an element of D* such that $\cap_{n=1}^{\infty} a^n D = (0)$. Compare the quotient fields of D[[X]] and (D[1/a])[[X]]. Notice that condition (3) of Theorem 29 says that the quotient fields must differ in this case and so the question is a meaningful one. Sheldon also has something to say about determining whether or not it can happen that for an element a of D, D[[X]] and (D[1/a])[[X]] can have the same quotient field. We shall devote the remainder of this section to Sheldon's results. The main theorem has a difficult proof and we must prepare the groundwork.

Let R be a ring with f a power series in R[[X]]. By the *support* of f, written supp (f), we shall mean the set of all nonnegative integers i such that the coefficient of X^i in f is different from zero.

We now state the main theorem of this section. We have chosen to omit what to us is the tedious part of the argument. Hopefully, the essence remains. As Sheldon remarks, some of the ideas for the proof were derived from a paper of R. Rivet [29].

THEOREM 30. Let D be an integral domain and let a be a nonzero element of D. If $\cap_{n=1}^{\infty} a^n D = (0)$, then the quotient field of D[[X/a]] has infinite transcendence degree over the quotient field of D[[X]].

Proof. We first choose a function f from the positive integers into the positive integers having the following properties:

(a) $f(i) > i$, $i = 1,2,\ldots,$

(b) $f(i+1) > f(i) + 1$, $i = 1,2,\ldots,$ and

(c) For each integer $N > 0$, there exists an integer $I > 0$ such that $i \geq I$ implies that $f(i+1) > N.f(i)$.

An example of such a function is the function $f(i) = 2^{2^{2^i}}$. For each positive integer m, denote by f^m the function f composed with itself m times. Now, define an infinite sequence of power series in D[[X]] as follows:

$$a_1 = \sum_{i=1}^{\infty} (X/a)^{f(i)}$$

$$a_2 = \sum_{i=1}^{\infty} (X/a)^{f^2(i)}$$

.

.

.

$$a_j = \sum_{i=1}^{\infty} (X/a)^{f^j(i)}$$

Note that supp (a_1) = image (f), and supp (a_2) = image $(f^2) \subsetneq$ image (f) = supp (a_1). In fact, supp $(a_{j+1}) \subsetneq$ supp (a_j) for each j. Also, note that by our choice of f, a_1 has very large "gaps" in its support and that each successive a_j has increasingly larger gaps since its support is a very small subset of the support of the preceding power series. We shall show that the set $\{a_j\}_{j=1}^{\infty}$ is algebraically independent over the quotient field of D[[X]].

Suppose that the set $\{a_j\}_{j=1}^{\infty}$ is algebraically dependent. Since we may clear of denominators, we may assume that there is a polynomial $P \in (D[[X]])\,[T_1,\ldots,T_j]$ such that $P(a_1,\ldots,a_J) = 0$. Consider the set of all power series of the form $a_1^{e(1)} \ldots a_J^{e(J)}$ corresponding to those monomials $T_1^{e(1)} \ldots T_J^{e(J)}$ which occur in P with a nonzero coefficient. In this notation, the algebraic dependency relation satisfied by the elements $a_1,\ldots,a_J$ becomes a linear dependency relation among the elements of M. Order the elements of M lexicographically using the exponents of a_1, then a_2, and so forth, to establish the ordering. Denote the largest term in this ordering by $m_0 = a_1^{e'(1)} \ldots a_J^{e'(J)}$. A lengthy calculation, *which we will omit*, shows that the set has the following "Special Large Gap Property," SLGP:

For each integer $n > 0$, there is an integer L such that every power from X^{L-n} through X^{L+n} has a zero coefficient in each of the elements of M except that the coefficient of X^L in m_0 is a^{-L}.

Denote by $m_0, m_1, \ldots, m_t$ the elements of M. Then SLGP is much easier to visualize in terms of the $(t+1) \times \infty$ matrix whose (r,s) entry is the coefficient of X^s in the power series m_{r-1}. Thus, the SLGP means that, for each n, there exists a $(t+1) \times (2n+1)$ submatrix of the following form:

$$\begin{array}{c} \\ m_0 \\ m_1 \\ \cdot \\ \cdot \\ \cdot \\ m_t \end{array} \begin{array}{c} \begin{array}{ccccc} X^{L-n} & X^{L-1} & X^{L} & X^{L+1} & X^{L+n} \end{array} \\ \left[\begin{array}{ccccc} 0,\ldots, & 0, & a^{-L}, & 0,\ldots, & 0 \\ 0,\ldots, & 0, & 0, & 0,\ldots, & 0 \\ \cdot & \cdot & \cdot & \cdot & \cdot \\ \cdot & \cdot & \cdot & \cdot & \cdot \\ \cdot & \cdot & \cdot & \cdot & \cdot \\ 0,\ldots, & 0, & 0, & 0,\ldots, & 0 \end{array}\right] \end{array}$$

We now proceed to derive a contradiction. Let

$$f_0m_0 + f_1m_1 + \ldots + f_tm_t = 0$$

be the linear dependence relation of $m_0, m_1, \ldots, m_t$ over D[[X]]. Then f_0 is a nonzero power series in D[[X]]. If the order of f_0 is k, let d be the initial coefficient of f_0. Pick any positive integer n greater than k and let L be the integer specified in the SLGP which corresponds to the value n. For $0 \leq i \leq t$, we make the following definitions:

m'_i = the sum of the terms of m_i of degree less than L-n

m''_i = the sum of the terms of m_i of degree greater than L+n.

Then m'_i is a polynomial in D[X/a] and m''_i is a power series in D[[X/a]]. With these definitions we may rewrite the original m'_i's thusly:

$$\begin{aligned} m_0 &= m'_0 + a^{-L}X^L + m''_0 \\ m_1 &= m'_1 + \qquad\qquad m''_1 \\ &\;\;\vdots \\ m_t &= m'_t + \qquad\qquad m''_t \end{aligned}$$

Set

$$c' = f_0m'_0 + f_1m'_1 + \ldots + f_tm'_t$$

and

$$c'' = f_0m'' + f_1m''_1 + \ldots + f_tm''_t$$

Then the linear dependence relation becomes

$$c' + f_0(a^{-L}X^L) + c'' = 0$$

Observe that each of the power series m''_i has all zero coefficients before the $(L+n)^{th}$-degree term and so the same can be said for c''.

Recall that $m'_0, m'_1, \ldots, m'_t$ are all elements of D[[X/a]] and each has no term after the $(L-n)^{th}$-degree term. Thus, if we multiply each one by a^{L-n}, it will become a polynomial in D[X]. Hence $(a^{L-n})c'$ is also in D[[X]]. If we multiply the equation above by a^{L-n}, we will get the equation

$$(a^{L-n})c' + f_9(a^{-n} X^L) + a^{L-n} c'' = 0$$

Now consider the coefficient of X^{L+k} in the left hand side of the equation. This coefficient must be zero since the whole expression is zero. On the other hand, it must be equal to the sum of the coefficients of X^{L+k} in each of the three terms in the expression. The last term has a zero coefficient since $L + k < L + n$. The first term is a power series in D[[X]] and so its coefficient is in D. We are forced to conclude that the middle term has its coefficient of X^{L+k} in D as well. But this coefficient is precisely da^{-n}. Hence, $d \in a^n D$. Now our original choice of n could have been replaced by any larger value without changing the conclusion. Therefore, d is a nonzero element of $\bigcap_{n=1}^{\infty} a^n D$. This contradiction proves that the set $\{a_j\}_{j=1}^{\infty}$ is algebraically independent.

Theorem 30 has the following interesting corollary.

THEOREM 31. Let D be a principal ideal domain with quotient field K. If $D_1 \subsetneq D_2$ are distinct subrings of K containing D, then the quotient field of $D_2[[X]]$ has infinite transcendence degree over the quotient field of $D_1[[X]]$. More concretely yet, if Z is the ring of integers and if R is an integral domain properly between Z and the field Q of rational numbers, then the quotient field of R[[X]] has infinite transcendence degree over the quotient field of Z[[X]].

Proof. It is well-known that since D is a principal ideal domain, so are D_1 and D_2. Pick any element $b/a \in D_2 \setminus D_1$, where a and b are relatively prime elements of D. Then there exist elements $y, z \in D$ such that $1 = ya + zb$. Thus, $1/a = y + (zb/a) \in D_2$ and clearly $1/a \notin D_1$. Unique factorization in D_1 implies that $\bigcap_{n=1}^{\infty} a^n D_1 = (0)$. By Theorem 30, the transcendence degree of the quotient field of $D_1[[X/a]]$ over the quotient field of $D_1[[X]]$ is infinite. But $D_1 \subseteq D_2$ and a is a unit of D_2. Consequently, $D_1[[X/a]] \subseteq D_2[[X]]$, and since the same containment relationship holds between their quotient fields, the result follows.

On the positive side of the ledger, (or on the negative side, depending upon your point of view), the next result illustrates that D[[X]] and (D[1/a])[[X]] can have the same quotient field. We first recall a definition.

Let $R \subseteq S$ be rings. The *conductor* of R in S is the set of all elements $r \in R$ such that $rS \subseteq R$. It is easy to show that the conductor is the largest ideal of R which is also an ideal of S.

THEOREM 32. Let D_1 and D_2 be integral domains with $D_1 \subseteq D_2$. If the conductor of D_1 in D_2 is nonzero, then the quotient fields of $D_1[[X]]$ and $D_2[[X]]$ are equal.

Proof. Let $d \in D_1^*$ belong to the conductor of D_1 in D_2. If f is any power series in $D_2[[X]]$, then $d\,f \in D_1[[X]]$ and so $f \in (D_1[[X]])_{D_1^*}$. Therefore, $D_2[[X]] \subseteq (D_1[[X]]_{D_1^*} \subseteq$ (quotient field of $D_1[[X]]$).

We conclude this section with an example which shows that both types of behavior can occur in the same domain.

Example. Let V be a rank two valuation domain with $P \subsetneq M$ the nonzero prime ideals of V and choose $a \in P \setminus (0)$ and $b \in M \setminus P$. Then $(V[1/b])[[X]]$ and $V[[X]]$ have the same quotient field by Theorem 32. To see this, one need only notice that P is the conductor of V in

$V[1/b]$. On the other hand, $\cap_{n=1}^{\infty} a^n V = (0)$ and so by Theorem 30 the transcendence degree of the quotient field of $V[[X/a]]$ over that of $V[[X]]$ is infinite. But $V[[X/a]] \subseteq (V[1/a])[[X]]$.

8. Integral Dependence in Power Series Rings

Suppose that $F \subseteq K$ are fields with K algebraic over F. Does it follow that $K((X))$ is algebraic over $F((X))$? This question is related to the following problem. Under the same hypotheses on F and K, is $K[[X]]$ integral over $F[[X]]$? More generally, we can ask, if $R \subseteq S$ are rings with S integral over R, is $S[[X]]$ integral over $R[[X]]$? Bourbaki has shown [7] that the analogous result does hold for polynomial rings and so there does seem reason to hope. However, since this is a chapter on pathology, the answer to each of these questions is "no."

Most of the results of this section are derived from the work of Gilmer [19]. Although [19] is a very nice paper, it treats these problems in a fashion more detailed than we require and consequently the material presented here is only a fraction of that given in [19].

DEFINITION. Let $R \subseteq S$ be rings. An element $s \in S$ is said to be *algebraic over* R if s satisfies a nonzero polynomial having coefficients in R.

Although being distinct concepts, the notions of being integral and of being algebraic are related as the following easy result illustrates.

THEOREM 33. Suppose that $R \subseteq S$ are rings and that R is a reduced ring. Then an element $s \in S$ is algebraic over R if and only if there exists an element $r \in R$, $r \neq 0$ such that rs is integral over R.

Proof. If s is algebraic over R, then $r_n s^n + \dots + r_1 s + r_0 = 0$ for some $r_n, \dots, r_0 \in R$ with $r_n \neq 0$. Then $(r_n s)^n + \dots + r_n^{n-2}(r_1 s) + r_n^{n-1} r_0 = 0$ and $r_n s$ is integral over R. Conversely, if rs is integral over R, then $(rs)^m + \dots + r_1(rs) + r_0 = 0$ for some $r_{m=1},\dots,r_0 \in R$. Since $r \neq 0$ and R is reduced, $r^m \neq 0$ and so s is algebraic over R.

We now come to a more substantial result.

THEOREM 34. Suppose that $R \subseteq S$ with S integral over R. Suppose further that $\{a_i\}_1^\infty$ is a sequence of elements of S such that for any n, the monic polynomial over $R[a_1,\dots,a_{n-1}]$ of minimal degree having a_n as a root has degree > n. Then the element $\Sigma_{i=1}^\infty a_i X^{i!}$ is not integral over R[[X]].

Proof. We denote by c the element $\Sigma_{i=1}^\infty a_i X^{i!}$. If k is a positive integer and $d_1,d_2,\dots,d_k \in R[[X]]$, we show that $c^k + d_1 d^{k-1} + \dots + d_k \neq 0$. Thus, let $p = \Sigma_{i=1}^k a_i X^{i!}$ and let $q = c - p$. Then q is divisible by $X^{(k+1)!}$. Hence, if s is a positive integer, then

$$c^s = (p+q)^s = p^s + (\text{terms divisible by } X^{(k+1)!})$$

We conclude that for any positive integer s, the coefficient of $X^{k(k!)}$ in c^s is the same as the coefficient of $X^{k(k!)}$ in p^s. Hence, the coefficient of $X^{k(k!)}$ in c^k has the form

$$a_k^k + r_1^{(k)} a_k^{k-1} + \dots + r^{(k)} \text{ where } r_i^{(k)} \in R[a_1,\dots,a_{k-1}]$$

We next consider the coefficient of $X^{k(k!)}$ in $d_i c^{k-i}$ for $0 < i < k$. If $c^{k-i} = \Sigma_{j=k-i}^\infty u_j^{(-i)} X^j$, we observe that for $k - i \leq j \leq k(k!)$, $u_j^{(k-i)}$ is a polynomial in $a_1,\dots,a_k$ with coefficinets in R having degree at most k - i in any a_j with $1 \leq j \leq k$. In particular, each such $u_j^{(k-i)}$ is a polynomial in a_k with coefficients in $R[a_1,\dots,a_{k-1}]$ of degree less than or equal to k - i in a_k. Therefore, the coefficient of $X^{k(k!)}$ in $d_i c^{k-i}$, for 0 < i <, has the form $\Sigma_{j=0}^{k-i} r_j^{(k-i)} a_k^j$, where $r_j^{(k-i)} \in R[a_1,\ldots,a_{k-1}]$. It then follows that the coefficient of $X^{k(k!)}$ in $c^k + d_1c^{k-1} + \ldots + d_k$ has the form $a_k^k + r_{k-1}a_k^{k-1} + \ldots + r_0$, where $r_i \in R[a_1,\ldots,a_{k-1}]$. From the hypothesis on a_k it then follows that the coefficient of $X^{k(k!)}$ in $c^k + d_1c^{k-1} + \ldots + d_k$ is non-zero, so that $c^k + d_1c^{k-1} + \ldots + d_k \neq 0$. Since $d_1,\ldots,d_k$ were arbitrary elements of R[[X]], we conclude that c is not integral over R[[X]].

We will have answered two of the questions posed at the beginning of the section as soon as we can find a pair of fields to which we can apply Theorem 24. This is easily done via the following result.

THEOREM 35. Suppose that L is a separable algebraic field extension of the field F. In order that L[[X]] be integral over F[[X]] it is necessary that the dimension of L over F be finite.

Proof. We seek to find a sequence satisfying the hypotheses of Theorem 34. Thus, take a_1 to be any element of $L \setminus F$. We proceed by induction. Being given $a_1,\ldots,a_r$ such that if $F_i = F(a_1,\ldots,a_i)$ for each i, then $[F_{i+1} : F_i] > i$ for $0 \leq i \leq r - 1$, and we observe that since L over F is infinite dimensional, we can find $b_1,\ldots,b_{r+1} \in L$ which are linearly independent over F_r. But $F_r(b_1,\ldots,b_{r+1})$ is separable over F_r and so we can choose an element $a_{r+1} \in F_r(b_1,\ldots,b_{r+1})$ such that $F_r(b_1,\ldots,b_{r+1}) = F_r(a_{r+1})$. By our choice of $b_1,\ldots,b_{r+1}$, $[F_r(a_{r+1}) : F_r] \geq r + 1$. This completes the inductive step.

Thus, for example, if Q is the field of rational numbers and if K is any finite dimensional algebraic extension of Q, then K[[X]] is not integral over Q[[X]]. We shall see that K((X)) is not algebraic over Q((X)), but this is a ways off.

For now, since the proof is so easy we proceed to a converse of Theorem 35. It follows directly from the following observation.

THEOREM 36. Suppose that $R \subseteq S$ are rings with S a finite R-module. Then S[[X]] is a finite R[[X]]-module. In particular, S[[X]] is integral over R[[X]].

Proof. If $s_1,\ldots,s_n$ generate S as a module over R, then it is easy to check that they also generate S[[X]] as a module over R[[X]].

We summarize our results in the next proposition.

THEOREM 37. Suppose that L is a separable algebraic extension field of the field F. In order that L[[X]] be integral over F[[X]] it is necessary and sufficient that $[L : F] < \infty$.

We turn now to an elementwise version of Theorem 37. Although we shall make further use of it ourselves, it is of independent interest.

THEOREM 38. Let $F \subseteq L$ be fields with L separable algebraic over F. Let $f \in L[[X]]$ and denote by K the subfield of L obtained by adjoining the coefficients of f to F. Then f is integral over F[[X]] if and only if $[K : F] < \infty$.

Proof. If $[K : F] < \infty$, then K[[X]] is integral over F[[X]] by Theorem 37. In particular, f is integral over F[[X]]. Conversely, suppose that f is integral over F[[X]]. Let K_0 be a normal closure of K and let G be the Galois group of K_0 over F. If $\sigma \in G$, then we can extend σ to an automorphism σ^* on $K_0[[X]]$ by defining $\sigma()k) = \sigma(k)$ for each $k \in K_0$ and $\sigma^*(X) = X$. Since f is integral over F[[X]], there exist elements $h_0,\ldots,h_{m-1} \in F[[X]]$ such that $f^m + h_{m-1} f^{m-1} + \ldots + h_0 = 0$. Therefore, for each element $\sigma \in G$, we have that

$$0 = \sigma^*(0) = \sum_{i=0}^{m} h_i[\sigma^*(f)]^i$$

Thus, $\sigma^*(f)$ is a room of $\Sigma_{i=0}^{m} h_i Y^i \in (F[[X]])[Y]$. Since F[[X]] is an integral domain, $\Sigma_{i=0}^{m} h_i Y^i$ has only finitely many roots, but by the definition of K and G, distinct elements of G will have distinct action on f. Thus, G is a finite set and $[K : F] < \infty$.

Lest the reader forget our goal, we are still intent on proving that L((X)) need not be algebraic over K((X)) even though L be algebraic over K. In case L is separable algebraic over K, Theorem 37 pinpoints exactly when L[[X]] is integral over K[[X]]. Now if we knew that L((X)) is algebraic over K((X)), our problem would be solved. Although such a theorem need not hold between a pair of domains and their quotient fields, we shall prove that in our set-up, it does. An inspection of the general situation reveals precisely what is needed and that is the content of Theorem 40. We require a preliminary result.

THEOREM 39. Let F be a subfield of the field K, let f be a nonzero element of F[[X]], and let g be a nonzero element of K[[X]]. Then the two fields F(coefficients of fg) and F(coefficients of g) are equal.

Proof. The containment F (coefficients of fg) $\subseteq$ (coefficients of g) is clear. Write $f = \Sigma_{i=m}^{\infty} a_i X^i$ and $g = \Sigma_{i=n}^{\infty} b_i X^i$ with $a_m \neq 0 \neq b_n$. Then $a_m^{-1}(a_m b_n) = b_n$ belongs to F (coefficients of fg) $= F_1$. By induction, assume that $b_n, \ldots, b_k \in F_1$ and set $h = b_n X^n + \ldots + b_k X^k$. The coefficients of fh are in F_1 and hence the coefficients of fg - fh differ from those of fg by elements of F_1. Thus, $F_1 \supseteq$ F (coefficients of fg - fh). But fg - fh = f(g-h) and by the first part of the argument $b_{k+1} \in$ F (coefficients of fg - fh). We conclude that f (coefficients of g) $\subseteq F_1$ = F (coefficients of fg).

THEOREM 40. Let K be a separable algebraic extension field of the field F. If $g \in K[[X]]$ is algebraic over F[[X]], then g is integral over F[[X]].

Proof. Since g is algebraic over F[[X]], there is a nonzero element $f \in F[[X]]$ such that fg is integral over F[[X]]. By Theorem 37, if $F_1 = F$ (coefficients of fg), then $[F_1 : F] < \infty$. But Theorem 39 asserts that $F_1 = F$ (coefficients of g) and so by the other part of Theorem 37, g is integral over F[[X]].

We come at last to the answer to the first question posed at the beginning of this section.

THEOREM 41. Let L be a separable algebraic field extension of the field F. Then L((X)) is algebraic over F((X)) if and only if L[[X]] is integral over F[[X]].

Proof. In general, it is immediate from the definitions that if D_1 is an integral domain with quotient field L_1 and if D_2 is an integral domain with quotient field L_2, then from D_2 integral over D_1 it follows that L_2 is algebraic over L_1.

If L((X)) is algebraic over F((X)), then let $g \in L[[X]]$. By hypothesis, g is algebraic over F((X)) and, by clearing denominators, g is algebraic over F[[X]]. From Theorem 40 we see that g is integral over F[[X]].

9. Power Series Rings Over von Neumann Regular Rings

We have here-to-fore said nothing about homological aspects of power series rings and in this section we shall partially remedy that. We will see, however, that very little is known and that much of what is known is negative. For example, although both the global dimension and the weak global dimension are well-behaved with respect to forming the polynomial ring, neither is well-behaved with respect to forming the power series ring. Nothing is known about the global dimension except the obvious, but the paper [8] of the author, E. Rutter, and J. Watkins does focus on the weak global dimension of R[[X]] in case R is von Neumann regular, that is, the case w.gl. dim R = 0. Most of the results of this section are drawn from [8]. This forces us to have the following standing assumption throughout this section:

R will always denote a von Neumann regular ring

We shall assume a certain familiarity with the arithmetic of such rings as well as a facility with the notion of weak global dimension.

Thus, suppose that R is a von Neumann regular ring. Then w.gl. dim R = 0. What conditions on R are necessary and sufficient in order that w.gl. dim R[[X]] = 1? Theorem 1 of [8] lists several conditions, each of which is equivalent to R[[X]] having weak global dimension one and we seek now to prove a version of that result. As always some terminology is needed.

By a *countable system of linear equations* over a ring S, we mean countably many linear equations with coefficients from S and involving a countable set of indeterminates where in each equation only a finite set of indeterminates occur with non-zero coefficient. A ring S is called $\aleph_0$*-algebraically compact* if each finitely solvable countable system of linear equations over S has a simultaneous solution.

We will call a ring S $\aleph_0$*-self-injective* provided each S-homomorphism ϕ from a countably generated ideal I of S into S is induced by multiplication by an element of S; that is, there exists an element $b \in S$ such that $\phi(a) = ba$ for each element $a \in I$.

Let S be a reduced ring and for elements $a,b \in S$, define $a \leq b$ if $ab = a^2$. This is equivalent to saying that b agrees with a on the support of a in each representation of S as a subdirect product of integral domains. Thus, $\leq$ is a partial order on S. Subsets V and W of S are said to be *disjoint* or *orthogonal* provided $xy = 0$ for all $x \in V$, $y \in W$. An element a of S is said to *separate* a subset V of S from a subset W of S in the partial order $\leq$ is a is an upper bound for V and is orthogonal to W. For any subset V of S, $\text{ann}(V) = \{y \in S \mid yx = 0$ for each $x \in V\}$.

Recall that a ring S is said to be a Bézout ring if each finitely generated ideal of S is principal.

There is another ring-theoretic notion we shall need. If S is a subring of a ring T, then S is said to be *ideally closed* in T if $IT \cap S = I$ for each ideal I of S.

Thusly armed, we state the main theorem of this section. Notice that some of the conditions are on R and some are on R[[X]].

THEOREM 42. Let R be a von Neumann regular. The following conditions are equivalent:

(1) R is $\aleph_0$-algebraically compact.

(2) Given any orthogonal pair S,T of countable subsets of R each consisting of elements which are pairwise orthogonal, there is an element which separates S from T in the partial order $\leq$.

(3) R is $\aleph_0$-self-injective.

(4) If R is a subring of a ring R_1, then R[[X]] is ideally closed in $R_1[[X]]$.

(5) R[[X]] is a Bézout ring, and

(6) the weak global dimension of R[[X]] equals 1.

Proof. (1) ⇒ (4): It clearly suffices to prove that $IR_1[[X]] \cap R[[X]] = I$ for each finitely generated ideal I of R[[X]]. Thus, let I be a finitely generated ideal of R[[X]] and suppose that $g \in IR_1[[X]] \cap R[[X]]$. Translated into equations this amounts to a countable system L of linear equations with coefficients in R and having a solution in R_1. Since R is von Neumann regular, R_1 is a faithfully flat R-algebra. Thus, by [7, p.36], each finite system of linear equations solvable in R_1 is solvable in R. Therefore, L is a countable system of linear equations in R which is finitely solvable in R. By assumption, L is solvable in R and retranslating we have that $g \in I$.

(4) ⇒ (5): Any von Neumann regular ring is a subring of a direct product of fields. Thus, we may so regard R, say $R \subseteq T$. Being given $f,g \in R[[X]]$, we must prove that (f,g)R[[X]] is principal. Now each element of R can be written as the product of a unit times an indempotent. Hence, we can write $f = \Sigma_{i=0}^{\infty} r_i e_i X^i$ and $g = \Sigma_{i=0}^{\infty} s_i e'_i X^i$,

where r_i, s_i are units of R and e_i, e'_i are idempotents of R. We claim that $(f,g)T[[X]] = hT[[X]]$, where $h = \Sigma_{i=0}^{\infty} (e_i + e'_i - e_i e'_i)X^i$. To see this, write $T = \prod_\alpha K_\alpha$, K_α a field. It is easily checked that $T[[X]] = \prod_\alpha K_\alpha[[X]]$. Now an ideal I of $K_\alpha[[X]]$ is generated by any power series q in I of minimal order. Therefore,

$$(f,g)T[[X]] = (f,g) \prod_\alpha (K_\alpha[[X]]$$

$$= ((\ldots, f_\alpha, \ldots), (\ldots, g_\alpha, \ldots))$$

$$= ((\ldots, h_\alpha, \ldots))$$

where $h_{\alpha_i} = 0$ if $f_{\alpha_i} = g_{\alpha_i} = 0$ and $h_{\alpha_i} = 1$, otherwise. We leave to the reader the task of showing that this observation substantiates the claim. But then, $h \in R[[X]]$ and by hypothesis

$$hR[[X]] = hT[[X]] \cap R[[X]] = (f,g)T[[X]] \cap R[[X]] = (f,g)R[[X]]$$

(5) ⇒ (2): Let $\{a_i\}_0^\infty$, $\{b_i\}_0^\infty$ be countable collections of elements of R such that $a_i a_j = 0 = b_i b_j$ for $i \neq j$ and $a_i b_j = 0$ for all i and j. Again, we can regard R as a subring of T, a direct product of fields. Now in T there is surely an element a such that $aa_i = a_i^2$ and $ab_i = 0$. Our problem is to separate $\{a_i\}_0^\infty$ from $\{b_i\}_0^\infty$ in R.

Write

$$f = a_0 + b_0X + a_1X^2 + b_1X^3 + \ldots$$

and

$$g = a_0^2 + a_1^2X^2 + a_2^2X^4 + \ldots$$

By hypothesis, $(F,g)R[[X]] = hR[[X]]$ for some power series h in R[[X]]. Thus, $f = f'h$ and $h = \alpha f + \beta g$, where f', α, and β belong to R[[X]]. Now, $g = af$ and so $h = \alpha f + \beta g = \alpha f + \beta af = (\alpha+\beta a)f'h$. Therefore, $[1-(\alpha+\beta a)f']h = 0$ and since $f \in hR[[X]]$, $[1-(\alpha+\beta a)f']f = 0$. By Theorem 10, the constant term $[1-(\alpha_0+\beta_0 a)f'_0]$ of $1 - (\alpha+\beta a)f'$ annihilates f. Hence, $[1-(\alpha_0+\beta_0 a)f'_0]a_i = 0 = [1-(\alpha_0+\beta_0 a_0)f'_0]b_i$. Write $f'_0 = ue$, u a unit of R and e an idempotent of R. Then $(1-e)f'_0 = 0$ and so $(1-e)a_i = (1-e)[1-(\alpha_0+\beta_0 a)f'_0]a_i = 0$. Similarly, $(1-e)b_i = 0$. Therefore, $ef = f$ and $eg = g$.

Set $f^* = (1-e) + f' = (1-e+ue) + f'_1X + f'_2X^2 + \ldots$, the f'_i's elements of R. Since $(1-e+ue)$ is a unit of R, f^* is a unit of R[[X]]. Moreover, $f^*h = (1-e)h + f'h = (1-e)(\alpha f+\beta g) + f'h = f'h = f$. Thus $fR[[X]] = hR[[X]] = (f,g)R[[X]]$ and so $g = \overline{g}f$ for some power series $\overline{g} \in R[[X]]$. But, if $\overline{g} = \overline{g}_0 + \overline{g}_1X + \ldots$, then

$$\overline{g}f = \overline{g}_0a_0 + \ldots + (\overline{g}_0a_i + \overline{g}_1b_{i-1} + \ldots + \overline{g}_{2i}a_0)X^{2i}$$

$$+ (\overline{g}_0b_i+\overline{g}_1a_i + \ldots + \overline{g}_{2i+1}a_0)X^{2i+1} + \ldots$$

Now, write $a_i = u_ie_i$ and $b_i = v_id_i$, where u_i, v_i are units and e_i, d_i are idempotents. Then

$$a_i^2 = e_ia_i^2 = e_i(\overline{g}_0a_i+\overline{g}_1b_{i-1} + \ldots + \overline{g}_{2i}a_0) = e_i\overline{g}_0a_i = \overline{g}_0a_i$$

and

$$0 = d_i \cdot 0 = d_i(\overline{g}_0b_i+\overline{g}_1a_i + \ldots + \overline{g}_{2i+1}a_0) = d_i\overline{g}_0b_i = \overline{g}_0b_i$$

It follows that $\overline{g}_0$ is an element of R which separates $\{a_i\}_0^\infty$ from $\{b_i\}_0^\infty$.

(2) ⇒ (3): Suppose that I is a countably generated ideal of R and that ϕ is an R-homomorphism from I into R. Since R is von Neumann regular an induction argument shows that we can assume that I is generated by a countable set $\{d_i\}_0^\infty$, where d_i is idempotent for all i and $d_id_j = 0$ for $i \neq j$. Write $a_i = \phi(d_i) = u_ie_i$, where u_i is a

unit of R and e_i is an idempotent. Set $f_i = d_i - e_i$. We would like to be able to separate $\{a_i\}_0^\infty$ from $\{f_i\}_0^\infty$. Note that $d_i a_i = d_i \phi(d_i) = \phi(d_i^2) = \phi(d_i) = a_i$ and so $u_i e_i d_i = u_i e_i$. Thus, $e_i d_i = e_i$ and so if $i \neq j$, then $f_i f_j = (d_i - e_i)(d_j - e_j) = (d_i - e_i d_i)(d_j - e_j d_j) = 0$. Also, $a_i a_j = d_i a_i d_j a_j = 0$ and $a_i f_j = a_i d_i (d_j - e_j d_j) = 0$. Finally, $a_i f_i = u_i e_i (d_i - e_i) = u_i (e_i d_i - e_i) = 0$. Therefore, by hypothesis $\{a_i\}_0^\infty$ and $\{f_i\}_0^\infty$ are separable, say ab $a \in R$. Thus, $aa_i = a_i^2$ and $af_i = 0$ for each i.

Now for each i, $(\phi(d_i) - ad_i)a_i = (a_i - ad_i)a_i = a_i^2 - aa_i = 0$ and so $(\phi(d_i) - ad_i)e_i = 0$. Also, $(\phi(d_i) - ad_i)f_i = (u_i e_i - ad_i)f_i = 0$. Therefore, $0 = (\phi(d_i) - ad_i)(f_i + e_i) = (\phi(d_i) - ad_i)d_i = \phi(d_i^2 - ad_i^2 = \phi(d_i) - ad_i$. Consequently, $\phi(d_i) = ad_i$ and ϕ is given by multiplication by the element a.

(3) ⇒ (1): Let

$$L_{.} = \sum a_{1i} Y_i = b_1$$

$$L_2 = \sum a_{2i} Y_i = b_2$$

.

.

.

be a countable system of linear equations for which any finite subsystem has a solution. For each subsystem the complete solution is obtained as a particular solution plus the complete solution to the associated homogeneous system. The complete solution of the associated homogeneous system is the kernel of a homomorphism between finitely generated free R-modules. Since R is von Neumann regular, the kernel of such a homomorphism is itself finitely generated. The values of Y_1, say, that occur as part of a solution of the associated homogeneous system are obtained by projecting the kernel onto the first coordinate. Thus, the occuring values of Y_1 are a finitely generated ideal of R. Since R is regular, this ideal is of the form R(1-e) for some idempotent element e of R. These remarks say that the values of Y_1 that occur in the complete solution of any finite subsystem are precisely the elements of the form $\alpha + R(1-e)$ where α occurs as the value of Y_1 in a particular solution. Moreover, if we choose the initial subsystems $L_1 = b_1$; $L_1 = b_1$; $L_2 = b_2$; etc., the first components of the solutions of the mth initial subsystem will be precisely the elements of the form $\alpha_m + R(1-e_m)$, where $(1-e_m)(1-e_i) = (1-e_i)$ whenever $m \leq i$.

Now set $\beta_m = \alpha_m e_m$. We claim that y occurs as the first component of a solution to $L_1 = b_1,\ldots,L_m = b_m$ if and only if $ye_m = \beta_m$. For if $y = \alpha_m + r(1-e_m)$, then $ye_m = \alpha_m e_m + r(1-e_m)e_m = \alpha_m e_m = \beta_m$ and if $ye_m = \beta_m = \alpha_m e_m$, then $y = ye_m + y(1-e_m) = \alpha_m e_m + y(1-e_m) = \alpha_m e_m + \alpha_m(1-e_m) + (y-\alpha_m)(1-e_m) = \alpha_m + (y-\alpha_m)(1-e_m)$. Therefore, in particular, we must have $\beta_k e_m = \beta_m$ if $k \geq m$.

Define a homomorphism ϕ from the ideal $I = (e_1,e_2,\ldots)$ into R by $\phi(e_i) = \beta_i$. We must see that ϕ is well-defined. Since $e_i e_j = e_i$ for $i \leq j$, $e_i R \subseteq e_j R$ for $i \leq j$ and so $I = \bigcup_{i=1}^{\infty} e_i R$ is a union of a chain of principal ideals. Suppose that $x \in I$, say $x = re_i = se_j$ for $i \leq j$. Then $0 = re_i - se_j = re_i e_j - se_j = (re_i - s)e_j$. Therefore, $0 = (re_i - s)e_j\beta_j = (re_i - s)\beta_j$ and hence $re_i\beta_j = s\beta_j$. Since $e_i\beta_j = \beta_i$, $r\beta_i = s\beta_j$ and so ϕ is well-defined.

By hypothesis, ϕ is induced by multiplication by an element β of R. But for each i, $\beta_i = \phi(e_i) = \beta e_i$. Thus, β occurs as the first component in some solution of each initial subsystem.

Applying this argument to Y_1, we can choose y_1 so that for each positive integer m there exists a solution γ_m to the initial subsystem $L_1 = b_1,\ldots,L_m = b_m$ such that the first component of γ_m is y_1. We use induction on n. Supposing that we have chosen $y_1,\ldots,y_n$ appropriately, consider the system $\Sigma_{i=n+1} a_{1i}Y_i = b_1 - \Sigma_{i=1}^{n} a_{1i}y_i$, $\Sigma i=1+n\ a_{2i}Y_i = b_2 - \Sigma_{i=1}^{n} a_{2i}y_i, \ldots$. Applying the above argument to Y_{n+1} yields an element $u_{n+1} \in R$ such that for each positive integer m there exists a solution γ_m to the subsystem $L_1 = b_1, L_2 = b_2, \ldots, L_m = b_m$ such that the first n + 1 components of γ_m are $y_1, \ldots y_{n+1}$, respectively.

We claim that $\gamma = (y_1, t_2, \ldots)$ is a solution to the original system, for consider the equation $L_m = b_m$. Only finitely many of the Y_j's are involved; say $Y_1, \ldots, Y_n$. By our construction, there exists a solution γ_m to the system $L_1 = b_1, \ldots, L_m = b_m$ such that $y_1, \ldots, y_n$ are the first n components of γ_m. Thus, γ satisfies the equation $L_m = b_m$.

We have only to prove the equivalence of condition (6) to one of the other conditions. We will prove that it is equivalent to (3) and for that we recall the following flatness criterion of Bourbaki [7, p.27].

If E is a finitely generated module over an arbitrary commutative ring S and if E is generated by $b_1, \ldots, b_n$, the E is a flat S-module if and only if the following condition is satisfied: If $a_1, \ldots, a_m \in E$ and if $s_1, \ldots, s_m \in S$ with $\Sigma_{i=1}^{m} s_i a_i = 0$, then there exist elements $t_{ij} \in S$, $i = 1, \ldots, m, j = 1, \ldots, n$ such that $a_i = \Sigma_{j=1}^{n} t_{ij} b_j$ for all i $0 = \Sigma_{i=1}^{m} s_i t_{ij}$ for all j.

(6) ⇒ (3): Suppose that I is a countably generated ideal of R and that ϕ is an R-homomorphism from I into R. As noted in the proof that (2) implies (3), we can assume that I is generated by a countable set $\{d_i\}_0^{\infty}$, where d_i is idempotent for each i and $d_i d_j = 0$ for $i \neq j$. Let $a_i = \phi(d_i)$ for each i and set $f = \Sigma_{i=0}^{\infty} d_i X^i$ and $g = \Sigma_{i=0}^{\infty} a_i X^{i+1} = X \cdot \Sigma_{i=0}^{\infty} a_i X^i$. Since s.gl. dim $R[[X]] = 1$, $(f,g)R[[X]]$ is a flat ideal. Therefore, the Bourbaki flatness criterion and the relation $(-g)f + fg = 0$ imply the existence of power series h, k, α, and $\beta \in R[[X]]$ such that

$$f = hf + kg \qquad (-g)h + f\alpha = 0 \qquad (*)$$

and

$$g = \alpha g + \beta g \qquad (-g)k + f\beta = 0 \qquad (**)$$

By equating coefficients of X^i in the first equation of (*) we obtain the equation $d_i = \Sigma_{\ell+m=i} h_\ell d_m + \Sigma_{\ell+m=i-1} k_\ell a_m$, where h_ℓ and k_ℓ denote the coefficient of X^ℓ in h and k, respectively. Multiplying this equation by d_i, we obtain the relation $d_i = h_0 d_i$. Equating coefficients of X^{i+1} in the first equation of (**) we obtain the equation $\Sigma_{\ell+m=i+1} d_\ell \alpha_m = \Sigma_{\ell+m=i} a_\ell h_m$ and multiplying this equation by d_i, we obtain the relation $\alpha_1 d_i = h_0 a_i$. Recalling that $d_i a_i = a_i$, we have that $a_i = d_i a_i = h_0 d_i a_i = \alpha_1 d_i d_i = \alpha_1 d_i$. We conclude that ϕ is given by multiplication by α_1.

(3) $\Rightarrow$ (6): We know that (3) is equivalent to (5) so that R[[X]] is a Bézout ring. Let $f \in R[[X]]$. By the flatness criterion, the principal ideal fR[[X]] is flat if and only if for each $\alpha \in R[[X]]$ with $\alpha f = 0$ there exists $\beta \in R[[Z]]$ such that $\beta f = f$ and $\alpha\beta = 0$. Let $I = (\alpha_i)R$ and $J = (f_i)R$ where α_i and f_i are the coefficients of α and f, respectively. Since R is von Neumann regular, $I \cap J = IJ = (0)$. Therefore, the projection map $\phi\colon I + J \to R$ via $\phi(i+j) = j$ is well-defined and by hypothesis, ϕ is given by multiplication by an element $b \in R$. It follows that $bf = f$ and that $b\alpha = 0$. Thus, R[[X]] is a Bézout ring in which all principal ideals are flat. Such a ring clearly has weak global dimension less than or equal to one. This completes the proof.

The property of being $\aleph_0$-algebraically compact is a stringent one to place on a von Neumann regular ring and therefore, one is left with the impression that the weak global dimension of R[[X]] usually exceeds one. Unfortunately, there is no information regarding the weak global dimension of R[[X]] in case R is not $\aleph_0$-algebraically compact and so we pose the following problem.

Question. Suppose that R is a von Neumann regular ring which is not $\aleph_0$-algebraically compact. What can be said about the weak global dimension of R[[X]]? In particular, can it be infinite?

We interject a remark here which may or may not be relevant. Recall from Theorem 20 that if a ring S is not an A-ring, then the Krull dimension of R[[X]] is infinite and from Theorem 23 that any A-ring has Noetherian prime spectrum. Now a von Neumann regular ring has Noetherian prime spectrum if and only if it is trivial, that is, a direct sum of fields [See Exercise 7]. Hence, by Theorems 19 and 20, the power series ring over any trivial von Neumann regular ring has Krull dimension one and the power series ring over any nontrivial von Neumann regular ring has infinite Krull dimension. Could one expect similar behavior of the weak global dimension?

Recall that a ring S is called *semihereditary* if each finitely generated ideal of S is projective. If S is semihereditary, then S has weak global dimension less than or equal to one. We now turn to determining necessary and sufficient conditions on R in order that R[[X]] be semihereditary. This turns out to be fairly easy modulo Theorem 42.

THEOREM 43. Let R be a von Neumann regular ring. The following conditions are equivalent.

(1) R is $\aleph_0$-self-injective and every countably generated ideal of R has a countably generated annihilator.,

(2) Every countable subset of R that forms a chain in the partial order $\leq$ has a least upper bound in R.,

(3) R[[X]] is semihereditary., and

(4) R[[X]] is a Bézout ring in which all principal ideals are projective.

Proof. Is is convenient to begin by establishing the following result.

CLAIM. Let $a_1 \leq a_2 \leq \ldots$ be a chain in R with $a_i = u_i a_i$, where u_i is a unit and e_i is an idempotent. Then a = ue, u, a unit, e, an idempotent, is a least upper bound for the chain $a_1 \leq a_2 \leq \ldots$ if and only if e is a least upper bound for $e_1 \leq e_2 \leq \ldots$ and $ue_i = u_i e_i = a_i$ for each i.

Proof. Suppose that a is a least upper bound for $a_1 \leq a_2 \leq \cdots$. Then for each i, $Re_i = Ra_i = Ra_i^2 \subseteq Ra = Re$ and so $ee_i = e_i$ for each i. Therefore, e is an upper bound for $e_1 \leq e_2 \leq \cdots$. If $f \leq e$ and $e \leq f$ for each i, then set $b = uf$. Now, $a_i a = u_i e_i ue = u_i e_i ue_i = a_i^2 = u_i e_i u_i e_i$ and so $u_i e_i = ue_i$. Thus, $ba_i = ufu_i e_i = ue_i u_i e_i = u_i e_i u_i e_i = a_i^2$ and $a_i \leq b$ for each i. But then $a \leq b$ from which it follows that $e \leq f$. Consequently, e is the least upper bound for $e_1 \leq e_2 \leq \cdots$. Conversely, we obviously have that $a \geq a_i$ for each i. If $b \geq a_i$ for each i, then $a_i(ab-a^2) = a_i ab - a_i a^2 = a_i^2 a - a_i^2 a = 0$, that is $(ab-a^2)$ belongs to the annihilator, $\mathrm{ann}((a_1,a_2,\ldots))$, of $(a_1,a_2,\ldots)$. We claim that $\mathrm{ann}((a_1,a_2,\ldots)) = \mathrm{ann}((e_1,e_2,\ldots)) = \mathrm{ann}(Re)$. The first equality is clear and it is also clear that if $xe = 0$, then $xe_i = 0$ for each i. If $xe_i = 0$ for each i, then $(1-x)e_i = e_i$ and since e is the least upper bound for $e_1 \leq e_2 \leq \ldots$, $(1-x) \geq e$. Therefore, $e - xe = e^2 = e$ and so $xe = 0$. It follows that $ab - a^2 \in \mathrm{ann}(re) \cap Re = (0)$. Consequently, $a \leq b$.

We now proceed with the proof of the theorem.

(1) ⇒ (2): Let $a_1 \leq a_2 \leq \cdots$ be a chain in R with $a_i = u_i e_i$, u_1, a unit, e_i, an idempotent. Then $e_1 \leq e_2 \leq \cdots$ and $a_j e_i = a_i$ if $i \leq j$. Define a homomorphism ϕ: $(e_1,e_2,\ldots) \to R$ by $\phi(e_i) = a_i$. Now ϕ is well-defined for if $e_i \leq e_j$, then $\phi)e_i) = \phi(e_i e_j) = e_i e_j = a_i$. By hypothesis, there exists an element $b \in R$ such that $be_i = a_i$ for each i.

We claim that the chain $e_1 \leq e_2 \leq \cdots$ has a least upper bound in R. To verify this, we first prove that $\mathrm{ann}((e_1,e_2,\ldots)) = (1-e)R$ for some idempotent element e of R. It is then easy to see that e is the least upper bound for $e_1 \leq e_2 \leq \cdots$. Since $(e_1,e_2,\ldots) \cap \mathrm{ann}((e_1,e_2,\ldots)) = (0)$, the projection map from $(e_1,e_2,\ldots) + \mathrm{ann}(e_1,e_2,\ldots))$ onto $\mathrm{ann}((e_1,e_2,\ldots))$ and orthogonal to $(e_1,e_2,\ldots)$ is well-defined.

By hypothesis, the map is induced by an element 1 - e and it follows readily that $ann((e_1,e_2,\ldots)) = (1-e)R$.

Now, set a = be, e the least upper bound of $e_1 \leq e_2 \leq \ldots$. Since $e_ie = e_i$, $a_i = e_ib = e_ia$. Thus, $(a_1,a_2,\ldots) = (e_1,e_2,\ldots) \subseteq Ra$ and consequently, $Re \subseteq Ra \subseteq Re$. If we write a = ue for some unit u of R, then $u_ie_i = a_i = ae_i = uee_i = ue_i$. It follows from the Claim that a is the least upper bound for $a_1 \leq a_2 \leq \ldots$.

(2) ⇒ (4): To prove that R[[X]] is a Bézout ring, it suffices, in light of Theorem 42, to prove that if $\{a_i\}_0^\infty$ and $\{b_i\}_0^\infty$ are countable subsets of R with $a_ia_j = b_ib_j = 0$ for $i \neq j$, and $a_ib_j = 0$ for all i and j, then there exists an element a of R with $aa_i = a_i^2$ and $ab_i = 0$ for each i. For each positive integer n, define $a'_n = \Sigma_{i=0}^n a_i$. Then $a'_ia'_j = a_i'^2$ for $i \leq j$ and hence, $a'_1 \leq a'_2 \leq \cdots$. By hypothesis, this chain has a least upper bound, say a, and this a will suffice.

Suppose next that $f = \Sigma_{i=0}^\infty b_iX^i$ is an element of R[[X]]. By Theorem 10, ann(fR[[X]]) = I[[X]], where $I = ann_R((b_0,b_1,\ldots))$. We can write $(b_0,b_1,\ldots) = (e_0,e_1,\ldots)$ where e_i is idempotent and $e_0 \leq e_1 \leq \ldots$. By hypotheses, $e_0 \leq e_1 \leq \ldots$ has a least upper bound e in R. It is easily verified that $(1-e)R = ann(eR) = ann(e_p,e_1,\ldots)$. Therefore, ann(fR[[X]]) = (1-e)R[[X]] and fR[[X]] is a projective ideal of R[[X]].

(3) ⇒ (4): That (4) implies (3) is obvious from the definitions and (3) implies (4) follows from Theorem 42 and the fact that a semi-hereditary ring has weak global dimension less than or equal to one.

(4) ⇒ (1): That R is $\aleph_0$-self-injective follows from Theorem 42. If $(a_0,a_1,\ldots)$ is a countably generated ideal of R, let f be the power series $f = a_0 + a_1X + \ldots$. Since fR[[X]] is a projective ideal, ann(fR[[X]] is generated by an idempotent of R[[X]].. But all idempotent elements of R[[X]] lie in R and so ann(fR[[X]] = eR[[X]], $d \in R$. Clearly, $Re = ann((a_0,a_1,\ldots))$. This completes the proof.

As we remarked before the proof of Theorem 43, if R[[X]] is semihereditary, then R[[X]] has weak global dimension less than or equal to one and this fact is obvious from Theorems 42 and 43. We conclude this section with an example of a Boolean ring B which satisfies the hypotheses of Theorem 42 but not those of Theorem 43. Thus, even for Boolean rings, the conditions of Theorems 42 and 43 are not equivalent. The example clearly illustrates the utility of the partial order we have been using to study R.

Example. Let Z be the ring of integers and let $R = (\prod_1^\infty (Z/2Z)/(\bigoplus_1^\infty (Z/2Z))$. R is clearly a Boolean ring and we claim that the weak global dimension of R[[X]] is one. By Theorem 42, it suffices to show that if $\{a_i\}_1^\infty$ and $\{b_i\}_1^\infty$ are countable subsets of R such that $a_i a_j = b_i b_j = 0$ for $i \neq j$ and $a_i b_i = 0$ for all i and j, then there exists an element $a \in R$ such that $aa_i = a_i$ and $ab_i = 0$ for all i. For each i, let A_i be a representative for a_i and B_i a representative for b_i. If $C \in \prod_1^\infty (Z/2Z)$, let supp(C) be the set of all positive integers i such that $C(i) = 1$, that is, the support of C. Now, let A be that element of $\prod_1^\infty (Z/2Z)$ whose support is the set $(\text{supp}(A_1) \setminus \text{supp}(B_1)) \cup (\text{supp}(A_2) \setminus (\text{supp}(B_1) \cup \text{supp}(B_2))) \cup \ldots$. Then for each i, $\text{supp}(A_i) \supseteq \text{supp}(A_i) \cap \text{supp}(A) \supseteq (\text{supp}(A_i) \setminus (\text{supp}(B_1) \cup \ldots \cup \text{supp}(B_i)) = (\text{supp}(A_i) \setminus \text{supp}(B_1) \cap \ldots \cap (\text{supp}(A_i) \setminus \text{supp}(B_i))$ and so $aa_i = a_i$, if a is the image of A. On the other hand, $\text{supp}(B_i) \cap A = \text{supp}(B_i) \cap [(\text{supp}(A_1) \setminus \text{supp}(B_1)) \cup \ldots \cup (\text{supp}(A_{i-1}) \setminus (\text{supp}(B_1) \cup \ldots \cup \text{supp}(B_{i-1}))]$. Therefore, since $b_i a_j = 0$, $\text{supp}(B_i) \cap \text{supp}(A_j)$ is finite and so $\text{supp}(B_i) \cap \text{supp}(A)$ is finite. It follows that $b_i a = 0$.

To prove that R[[X]] is not semihereditary we need only exhibit a countable subset of R which has no least upper bound in R. To that end, let $\{S_i\}_1^\infty$ be a collection of pairwise disjoint infinite subsets of the set of positive integers. For each i, let A_i be the element of $\prod_1^\infty(Z/2Z)$ whose support is S_i and let a_i be its image in R. Suppose that a is an upper bound for $\{a_i\}_1^\infty$ and let A be a preimage of a. For each positive integer i, choose an $x_i \in \text{supp}(A) \cap \text{supp}(A_i)$ and let $S = \text{supp}(A) \setminus \{x_i\}_1^\infty$. If B is that element of $\prod_1^\infty(Z/2Z)$ whose support is S and if b is the image of B, then $a_i \leq b$ for each i and $b < a$. Therefore, $\{a_i\}_1^\infty$ has no least upper bound.

10. The Cancellation Problem for Power Series Rings

In [22] M. Hochster gave an example of rings A and B for which $A[X] \cong B[Y]$, but $A \not\cong B$. We present here a procedure due to A. Magid for extending A and B to obtain rings $\hat{A}$ and $\hat{B}$ such that $\hat{A}[[X]] \cong \hat{B}[[X]]$, but $\hat{A} \not\cong \hat{B}$.

First recall that in Hochster's example $A = S_R(P)$ is the symmetric algebra of a projective R-module P, where R is the coordinate ring of the real 2-sphere and P is the projective R-module corresponding to the tangent bundle. Thus $P \oplus R \cong R^{(3)}$, while $P \not\cong R^{(2)}$. Now $S_R(P) = \bigoplus_{i\geq 0} S_R^i(P)$, where $S_R^i(P)$ is the ith symmetric power of P. We define $\hat{A}$ to be the "complete symmetric algebra" $\hat{S}_R(P) = \prod_{i\geq 0} S_R^i(P)$, with the obvious multiplication. In Hochster's example, $B = S_R(R^{(2)}) \cong R[X_1,X_2]$, and we likewise define $\hat{B} = \hat{S}_R(R^{(2)})$.

We leave it to the reader as an exercise to show that if M is an R-module (R any ring), then $\hat{S}_R(M \oplus R) \cong \hat{S}_R(M)[[X]]$. Then we have $\hat{B} \cong R[[X_1,X_2]]$. Moreover, $\hat{A}[[X]] = \hat{S}_R(P)[[X]] \cong \hat{S}_R(P \oplus R) \cong \hat{S}_R(R^{(2)} \oplus R) \cong \hat{S}_R(R^{(2)})[[Y]] = \hat{B}[[Y]]$.

Now suppose $\hat{A} \cong \hat{B}$. Let $\hat{I} = \prod_{i \geq 1} S_R^i(P)$ and $\hat{J} = \prod_{i \geq 1} S_R^i(R^{(2)})$. It is easy to see that $\hat{I}$, $\hat{J}$ are contained in the Jacobson radicals of $\hat{A}$, $\hat{B}$, respectively. And since $\hat{A}/\hat{I} \cong R \cong \hat{B}/\hat{J}$ and R has Jacobson radical zero, $\hat{I}$, $\hat{J}$, respectively, are the Jacobson radicals of $\hat{A}$ and $\hat{B}$. From the isomorphism $\hat{A} \cong \hat{B}$, we then obtain an isomorphism of associated graded rings $gr_{\hat{I}}(\hat{A}) \cong gr_{\hat{J}}(\hat{B})$, i.e. $\bigoplus_{i \geq 0} \hat{I}^i / \hat{I}^{i+1} \cong \bigoplus_{i \geq 0} \hat{J}^i / \hat{J}^{i+1}$. (See [6, Ch.10] for a nice treatment of graded rings.) But $\hat{I}^i / \hat{I}^{i+1} = S_R^i(P)$ and $\hat{J}^i / \hat{J}^{i+1} = S_R^i(R^2)$, so we get the contradiction $S_R(P) \cong S_R(R^2)$, i.e. $A \cong B$. Therefore $\hat{A} \not\cong \hat{B}$.

We close this chapter with some remarks about groups acting on power series rings. Let R be a commutative ring with X an indeterminate and let G be a group of R-automorphisms acting on R[[X]]. Denote by $(R[[X]])^G$ the fixed ring of G--that is,

$$(R[[X]]^G = \{h \in R[[X]] \mid \phi(h) = h \text{ for all } \phi \in G\}$$

Motivated by the polynomial case, we ask, "If G is a finite group, does there exist a power series f in R[[X]] such that $(R[[X]])^G = R[[f]]$?"

This problem was first studied by Samuel in [13], wherein it was shown, among other things, that the answer can be "no" if R contains nontrivial divisors of zero. Subsequent research of O'Malley [26] and of J. B. Catillon [9], [10] culminated in the following decisive theorem of O'Malley.

Theorem ([27, Theorem 1]). Let D be an integral domain with X an indeterminate and let G be a finite group of D-automorphisms acting on D[[X]]. If $g = \prod_{\phi \in G} \phi(X)$, then $(D[[X]])^G = D[[f]]$.

In [11], Castillon has tackled the analogous problem when the group is assumed to be infinite. Precisely, the following result is stated.

Let D be an integral domain with G an infinite group of D-automorphisms acting on D[[X]]. Then $(D[[X]])^G = D$. In our mind however, the proof has a serious gap and until it can be removed, the status of the result remains conjectural.

For a treatment of automorphisms of power series rings, see the next chapter.

Exercises

1. Let V be a rank one nondiscrete valuation domain having maximal ideal M. Prove that M is not an SFT-ideal and therefore that V is not an A-ring.
2. Let D be an integral domain with quotient field K. If α is an element of $K \setminus D$ with α almost integral over D, prove that the quotient field of D[[X]] and $(D[\alpha])[[X]]$ are equal.
3. Let I be an ideal of the ring R. Prove that $IR[[X]] = \{f \in R[[X]] \mid$ there exist elements $a_1, \ldots, a_n \in I$ such that $c(f) \subseteq (a_1, \ldots, a_n)\}$.
4. *Prove:* If $I[[X]] = IR[[X]]$ for each ideal I of a ring R, then R is Noetherian.
5. Suppose that R is a ring of dimension zero. Prove that the following conditions are equivalent:
   (1) $\dim R[[X]] \neq 1$.
   (2) $\dim R[[X]] = \infty$.
   (3) There exists a maximal ideal M of R such that $M[[X]] \neq \sqrt{MR[[X]]}$.
   Hint. Use Theorem 5.
6. Give an example of an integral domain D containing the ring Z of integers such that the quotient field of D[[X]] is algebraic over the quotient field of Z[[X]]. Conclude that Theorem 30 need not hold for arbitrary domains containing a principal ideal domain. (*Hint.* Any module-finite extension will do.)
7. Prove that a von Neumann regular ring has Noetherian prime spectrum if and only if it is a direct sum of fields.
8. Let R be a ring. Prove that each idempotent element of R[[X]] belongs to R.

CHAPTER 3

R-HOMOMORPHISMS OF $R[[X_1,...,X_n]]$

If R is a ring, then R[X] may be viewed as the free R-algebra on a single generator and thus, given any R-algebra S, there exist R-algebra homomorphisms from R[X] into S. Moreover, any such homomorphism ϕ is completely determined by its action on X. That is to say, if $\phi(X) = s$, then $\phi(a_0+a_1X+\cdots+a_nX^n) = a_0 + a_1s + \cdots + a_ns^n$ for each polynomial $a_0 + a_1X + \cdots + a_nX^n$ of R[X]. Thus, ϕ is merely "substitution by s." Since these remarks apply equally well to the R-algebra R[X] itself, they can be utilized to determine necessary and sufficient conditions on the image of X in order that an R-endomorphism be surjective or injective. In this manner, one computes the R-automorphisms of R[X].

To what extent do the above remarks apply to the power series situation? For example, what are the R-automorphisms of R[[X]]? A much more fundamental question is, given a power series f, does there exist an R-endomorphism ϕ of R[[X]] such that $\phi(X) = f$? Moreover, if such a morphism exists, is it unique? Now f could, of course, be a polynomial and then there certainly does exist an R[X] endomorphism ψ mapping X to f. Since R[[X]] is a completion of R[X] with respect to a certain topology, we are really asking whether or not we can lift ψ to a map on the completion. As often happens with lifting problems, the answer is "not always." For example, we will see that even in case R is the ring of integers, there does not exist an R-endomorphism ϕ of $R[[X]]$ such that $\phi(X) = 1 + X$. Intuitively, a difficulty lies in trying to determine where to map the power series $1 + X + X^2 + \cdots$.

In this chapter we shall investigate these problems in detail. It turns out that by treating them in greater generality we shall achieve not only a better understanding, but also a simplicity of argument. Thus, we are led to deviating from our customary policy of only considering the single variable case. Our conscience is clear in this matter for considering homomorphisms in the multi-variable case is far more natural.

11. Preliminaries and the Theorem of Eakin-Sathaye

In this section we prepare the groundwork for the remainder of the chapter. We shall fix the terminology to be used throughout the chapter as well as introduce the basic topological ideas which we shall need. The section concludes with a beautiful theorem of P. Eakin and A. Sathaye.

Dealing with topological considerations always presents one with the task of deciding how much detail to include and how much to omit. We shall be pragmatic in our approach and opt to omit most details. In fact, we shall only sketch the results we shall need and leave to the reader the chore of supplying the proof of those theorems with which he feels uncomfortable. Atiyah-MacDonald [6] has a concise chapter on completions, while Bourbaki's treatment [7] is, as usual, thorough and encyclopedic. There is also valuable material to be found in Zariski-Samuel [33,34], portions of which we shall include ourselves.

Let T be a commutative ring with I an ideal of T. By taking $\{I^n\}_{n=1}^{\infty}$ to be a fundamental system of neighborhoods of 0 in T, we endow T with a topology which is compatible with the ring structure on T. Thus, T is a topological ring and this topology is called the I-adic topology on T. The I-adic topology is Hausdorff if and only if $\cap_{n=1}^{\infty} I^n = (0)$.

If the topology is Hausdorff, it is, in fact, a metric topology and so we may speak of a completion of T. In this case T, being a topological ring, has more structure than an ordinary metric space and so when we speak of a completion of T we mean also that such a completion must be a topological ring and that it must stand in nice relation to T. More precisely, if T* is a completion of T, then T* must be a complete Hausdorff topological ring, T must be a unitary subring of T*, the topology on T is the relative topology it inherits from T*, and each element of T* is the limit of a Cauchy sequence of elements of T. One can show that such a completion does exist and that it is unique up to topological isomorphism.

Suppose now that R is a ring and that X_1, ..., X_m are indeterminates. The ideal $(X_1,\ldots,X_m)R[X_1,\ldots,X_m]$ of the ring $R[X_1,\ldots,X_m]$ induces a Hausdorff topological structure on $R[X_1,\ldots,X_m]$ and it is easy to prove that $R[[X_1,\ldots,X_m]]$ topologized with respect to its $(X_1,\ldots,X_m)R[[X_1,\ldots,X_m]]$-adic topology, is a completion.

As indicated previously, if S is any R-algebra, then there exist R-homomorphisms from $R[X_1,\ldots,X_m]$ into S and such R-homomorphisms are completely determined by their action on the set $\{X_1,\ldots, X_m\}$. Applying this remark to the R-algebra $R[[X_1,\ldots,X_m]]$, we know that given power series f_1, ..., $f_m \in R[[X_1,\ldots,X_m]]$, there exists an R-homomorphism ϕ from $R[X_1,\ldots,X_m]$ into $R[[X_1,\ldots,X_m]]$ such that $\phi(X_i) = f_i$ for each i. Can ϕ be lifted to the completion $R[[X_1,\ldots, X_m]]$--that is, does there exist an R-endomorphism ϕ^* of $R[[X_1,\ldots, X_m]]$ such that $\phi^*(X_i) = f_i$ for each i? If ϕ can be lifted, is the lifting unique? It is questions of this kind which we shall treat in this chapter. Although our notation does not indicate it, we must emphasize that our morphisms are R-algebra homomorphisms and, as such, must both preserve the ring structure as well as be R-linear.

We begin with a result from [34] which shows that if each of the target power series has constant term zero, then morphisms can always be lifted. In fact, as stated earlier, greater generality produces greater simplicity.

THEOREM 44. Let $X_1, \ldots, X_m, Y_1, \ldots, Y_n$ be indeterminates and let $f_1, \ldots, f_m \in R[[Y_1,\ldots,Y_n]]$ be power series each of which has constant term zero. Then there exists an R-homomorphism $\phi\colon R[[X_1,\ldots,X_m]] \to R[[Y_1,\ldots,Y_n]]$ such that $\phi(X_i) = f_i$ for each i. Moreover, regarding $R[[X_1,\ldots,X_m]]$ in the $(X_1,\ldots,X_m)$-adic topology and $R[[Y_1,\ldots,Y_n]]$ in the $(Y_1,\ldots,Y_n)$-adic topology, ϕ is continuous.

Proof. Let $g \in R[[X_1,\ldots,X_m]]$ and write $g = g_0 + g_1 + \cdots$ as a sum of forms. That is, g_i is a homogeneous polynomial in the variables $X_1, \ldots, X_m$ of degree i for $i \geq 0$. Thus, $g_i(f_1,\ldots,f_m)$, being a finite sum of power series, is again a power series. Moreover, since each f_i has constant term zero, the order of $g_i(f_1,\ldots,f_m)$ is greater than or equal to i for all i. Now the order of $g_i(f_1,\ldots,f_m)$ can exceed i, but there can be only finitely many forms g_j for which the order of $g_j(f_1,\ldots,f_m)$ is equal to the order of $g_i(f_1,\ldots,f_m)$. It follows that no "infinite summing" is necessary and that $g(f_1,\ldots,f_m)$ is indeed a power series in $R[[Y_1,\ldots,Y_n]]$. Define $\phi(g) = g(f_1,\ldots,f_m)$ to be this *substitution* of f_i for X_i and note that $\phi(g) = \lim_{i\to\infty}(g^{(i)}(f_1,\ldots,f_m))$, where $g^{(i)} = g_0 + g_1 + \cdots + g_i$ for all i. Here, the limit is being taken in the $(Y_1,\ldots,Y_n)$-adic topology on $R[[Y_1,\ldots,Y_n]]$. It is tedious, but not difficult, to prove that ϕ is indeed an R-homomorphism between $R[[X_1,\ldots,X_m]]$ and $R[[Y_1,\ldots,Y_n]]$ and it is clear that $\phi(x_i) = f_i$ for $1 \leq i \leq m$.

As for the continuity assertion, it is sufficient to prove that ϕ is continuous at zero and thus we have only to prove that the inverse image of a basic neighborhood of zero is open. For this, it suffices to show that given a positive integer s there exists a positive integer t such that $\phi((X_1,\ldots,X_m)^t) \subseteq (Y_1,\ldots,Y_n)^s$. Since each f_i has constant term zero, the order of $\phi(\alpha)$ is at least as big as the order of α for each $\alpha \in R[[X_1,\ldots,X_m]]$. These remarks substantiate the claim.

Theorem 44 is a fundamental result we shall often use. It is, however, the constant term zero case and in order to treat the case of nonzero constant term, much preliminary work is required. The principal tool we shall need is an elegant and powerful theorem of P. Eakin and A. Sathaye and to this we now turn our attention. We will split the proof of the theorem into three components.

THEOREM 45. Let $X_1, \ldots, X_m, Y_1, \ldots, Y_n$ be indeterminates over the ring R. Assume that σ: $R[[X_1,\ldots,X_m]] \to R[[Y_1,\ldots,Y_n]]$ is an R-homomorphism such that $\sigma(X_i) = c_i + f_i$, $c_i \in R$, $f_i \in (Y_1, \ldots,Y_n)$ for $1 \le i \le m$. For any subset $\{j_1,\ldots,j_s\}$ of $\{1,2,\ldots,m\}$, there exists an R-homomorphism σ: $R[[X_{j_1},\ldots,X_{j_s}]] \to R$ such that $\phi(X_{j_i}) = c_{j_i}$ for $1 \le i \le s$.

Proof. Let σ^* denote the restriction of σ to $R[[X_{j_i},\ldots,X_{j_s}]]$ and let τ be the R-homomorphism from $R[[Y_1,\ldots,Y_n]]$ onto R sending Y_i to 0 for all i. Then $\tau \circ \sigma^*$ is an R-homomorphism having the desired properties.

THEOREM 46. Let $X_1, \ldots, X_m$ be indeterminates over R with $c_1, \ldots, c_m$ elements of R. Suppose that there exists an R-homomorphism σ from $R[[X_1,\ldots,X_m]]$ onto R such that $\sigma(X_i) = c_i$ for $1 \le i \le m$. Then there exists an R-endomorphism τ of $R[[X_1,\ldots,X_m]]$ such that $\tau(X_i) = X_i + c_i$ for $1 \le i \le m$.

Proof. The trick is to introduce new indeterminates in order to give us room to operate. Thus, let $Z_1, \ldots, Z_m$ be indeterminates and define an R-homomorphism ϕ: $R[[Z_1,\ldots,Z_m]] \to R[[Z_1,\ldots,Z_m,X_1,\ldots,X_m]]$ by $\phi(Z_i) = X_i + Z_i$ for $1 \le i \le m$. Such a morphism does exist by Theorem 44. We wish now to define an R-homomorphism ψ from $R[[Z_1,\ldots,Z_m,X_1,\ldots,X_m]]$ to $R[[Z_1,\ldots,Z_m]]$ whose action on elements of $R[[X_1,\ldots,X_m]]$ is that of σ. Thus, we suggestively and abusively write elements of $R[[Z_1,\ldots,Z_m,X_1,\ldots,X_m]] = (R[[X_1,\ldots,X_m]])[[Z_1,\ldots,Z_m]]$ in the form $\Sigma_{i=0}^{\infty} g_i Z^i$, where $g_i \in R[[X_1,\ldots,X_m]]$ and Z^i denotes a form of degree i in the variables $Z_1, \ldots, Z_m$. Then define $\psi(\Sigma_{i=0}^{\infty} g_i Z^i) = \Sigma_{i=0}^{\infty} \sigma(g_i) Z^i$. This morphism is treated more formally in Exercise 1 where the reader is also asked to prove that it is indeed an R-homomorphism. Now the composition $\psi \circ \phi$ is an R-endomorphism of $R[[Z_1,\ldots,Z_m]]$ such that $(\psi \circ \phi)(Z_i) = c_i + Z_i$. But now the R-isomorphism between $R[[Z_1,\ldots,Z_m]]$ and $R[[X_1,\ldots,X_m]]$ which "changes the name" of Z_i to X_i, can be applied to yield the desired R-endomorphism of $R[[X_1,\ldots,X_m]]$.

THEOREM 47. Let $X_1, \ldots, X_m$ be indeterminates with $c_1, \ldots, c_m$ elements of R. Suppose that there exists an R-endomorphism τ of $R[[X_1,\ldots,X_m]]$ such that $\tau(X_i) = X_i + c_i$ for $1 \le i \le m$. Then τ is an automorphism.

Proof. Let σ be the R-automorphism of $R[[X_1,\ldots,X_m]]$ such that $\sigma(X_i) = -X_i$ for $1 \le i \le m$. Then the R-endomorphism $\sigma \circ \tau \circ \sigma \circ \tau$ of $R[[X_1,\ldots,X_m]]$ sends X_i to X_i for $1 \le i \le m$. Now any such morphism is continuous since it carries the ideal $(X_1,\ldots,X_m)$ into itself and any two such morphisms must agree on polynomials. But then, being continuous, they also must agree on limits of polynomials--that is, on power series. It follows that $\sigma \circ \tau \circ \sigma \circ \tau$ is the identity morphism and that τ is an automorphism.

Combining Theorems 45, 46, and 47 yields the promised result.

THEOREM 48 (Eakin-Sathaye) [12]. Let $X_1, \ldots, X_m, Y_1, \ldots, Y_n$ be indeterminates of the ring R. Assume that σ: $R[[X_1,\ldots,X_m]] \to R[[Y_1,\ldots,Y_n]]$ is an R-homomorphism and that for each i, $\sigma(X_i) = c_i + f_i$, where $c_i \in R$ and $f_i \in (Y_1,\ldots,Y_n)$. Then there exists an R-automorphism τ of $R[[X_1,\ldots,X_m]]$ such that $\tau(X_i) = X_i + c_i$ for each i.

The utility of Theorem 48 may not be obvious at first glance, but we shall devote the entire next section to its consequences.

12. Applications of the Theorem of Eakin-Sathaye

This section is devoted to deriving a host of conclusions from Theorem 48. In the process we will go a long way toward solving the problems mentioned at the beginning of the chapter.

We do not yet know when or if an R-homomorphism can exist between two power series rings for certain choices of the targets, but we can settle the uniqueness question at once.

THEOREM 49. Let R be a ring with $X_1, \ldots, X_m, Y_1, \ldots, Y_n$ indeterminates. Suppose that there exists an R-homomorphism ϕ: $R[[X_1,\ldots,X_m]] \to R[[Y_1,\ldots,Y_n]]$ such that for $1 \le i \le m$, $\phi(X_i) = c_i + f_i$, where $c_i \in R$ and $f_i \in (Y_1,\ldots,Y_n)$. Then ϕ is the only such R-homomorphism.

Proof. Let ϕ_1 and ϕ_2 be two such R-homomorphisms. By Theorem 48, there exists an R-automorphism τ of $R[[X_1,\ldots,X_m]]$ such that $\tau(X_i) = X_i + c_i$ for $1 \le i \le m$. Thus, $\tau^{-1}(X_i) = X_i - c_i$ for $1 \le i \le m$. It is clear that $\tau^{-1}(R[X_1,\ldots,X_m]) \subseteq R[X_1,\ldots,X_m]$ and moreover that ϕ_1 and ϕ_2 agree on $R[X_1,\ldots,X_m]$. Now,

$$(\phi_1 \circ \tau^{-1})(X_i) = \phi_1(X_i - c_i) = f_i$$

for $1 \le i \le m$ and the same is true of $\phi_2 \circ \tau^{-1}$. As in the proof of Theorem 44, $\phi_1 \circ \tau^{-1}$ and $\phi_2 \circ \tau^{-1}$ are continuous and therefore are continuous R-homomorphisms from $R[[X_1,\ldots,X_m]]$ into $R[[Y_1,\ldots,Y_n]]$ which agree on the dense subspace $R[X_1,\ldots,X_m]$ of $R[[X_1,\ldots,X_m]]$. Thus, they agree on limits of polynomials--that is, on $R[[X_1,\ldots,X_m]]$. Since τ^{-1} is an R-automorphism, $\phi_1 = \phi_2$.

Thus, *if* there is an R-homomorphism between $R[[X_1,\ldots,X_m]]$ and $R[[Y_1,\ldots,Y_n]]$, where the images of the X_i's have been specified, there is only one. The existence question is another story. It is best tackled by means of another idea from [12].

THEOREM 50. Let R be a ring. Set

$I_c(R) = \{a \in R \mid$ there exists an R-endomorphism σ of $R[[X_1]]$ such that $\sigma(X_1) = a + X_1\}$

$I_1(R) = \{a \in R \mid$ there exists an R-homomorphism $\sigma\colon\ R[[X_1]] \to R$ such that $\sigma(X_1) = a\}$

$I_2(R) = \{a \in R \mid$ there exists an R-homomorphism $\sigma\colon R[[X_1,\ldots,X_m]] \to R[[Y_1,\ldots,Y_n]]$, $X_1, \ldots, X_m, Y_1, \ldots, Y_n$ indeterminates of R, such that $\sigma(X_i) = a + f$ for some $X_i \in \{X_1,\ldots,X_m\}$ and $f \in (Y_1,\ldots,Y_n)\}$

Then $I_c(R) = I_1(R) = I_2(R)$ is an ideal of R containing the nilradical and contained in the Jacobson radical.

Proof. If $a \in I_c(R)$, then $a \in I_1(R)$ by Theorem 45. If $a \in I_1(R)$, then $a \in I_2(R)$ by Theorem 46. If $a \in I_2(R)$, then $a \in I_c(R)$ by Theorems 45 and 46. Therefore, $I_c(R) = I_1(R) = I_2(R)$.

To prove that I_c is an ideal of R, it suffices to prove that $ra + sb \in I_c$ for arbitrary elements $a,b \in I_c$ and $r,s \in R$. Again we introduce new indeterminates. Thus, let X, Y, and Z be indeterminates. Since $I_c(R) = I_2(R)$, it follows from Theorem 45 that there exists an R-homomorphism ϕ from $R[[X,Y]]$ onto R such that $\phi(X) = a$ and $\phi(Y) = b$. Let σ be the R-homomorphism from $R[[Z]]$ into $R[[X,Y]]$ given by $\sigma(Z) = rX + sY$. Then $\phi \circ \sigma$ is an R-homomorphism from $R[[Z]]$ onto R such that $(\phi \circ \sigma)(Z) = ra + sb$. By the first part of the theorem, $ra + sb \in I_c(R)$.

Finally, let $\alpha \in I_c(R)$. Thus, there exists an R-automorphism σ of $R[[X]]$ such that $\sigma(X) = a + X$. Since X belongs to the Jacobson radical of $R[[X]]$ so does $\sigma(X) = a + X$. It then follows that a belongs to the Jacobson radical of R (cf. Exercise 3). That $I_c(R)$ contains the nilradical of R is also left for the reader as Exercise 4.

We now prove a result which is a converse form of Theorem 50.

THEOREM 51. Let $X_1, \ldots, X_m, Y_1, \ldots, Y_n$ be indeterminates of R. Let $a_1, \ldots, a_s \in I_c(R)$ where $s \leq m$ and let $f_1, \ldots, f_s$ be elements of $(Y_1,\ldots,Y_n)R[[Y_1,\ldots,Y_n]]$. Then there exists an R-homomorphism τ: $R[[X_1,\ldots,X_s]] \to R[[Y_1,\ldots,Y_n]]$ such that $\tau(X_i) = a_i + f_i$ for $1 \leq i \leq s$.

Proof. By Theorem 50 and Theorem 45 there exists an R-homomorphism σ: $R[[X_1,\ldots,X_s]] \to R$ such that $\sigma(X_i) = a_i$ for $1 \leq i \leq s$. We may then extend σ to an R-homomorphism, abusively called σ, of $R[[X_1,\ldots,X_m]]$ by setting $\sigma(X_i) = 0$ for $s < i \leq m$. (This may be seen by regarding $R[[X_1,\ldots,X_m]]$ as $(R[[X_1,\ldots,X_s]])[[X_{s+1},\ldots,X_m]]$. The new σ is merely the result of first projecting onto the constant term, which lies in $R[[X_1,\ldots,X_s]]$, and then applying the old σ.)

Again, we need new indeterminates. Thus, let $Z_1, \ldots, Z_s$ be indeterminates and define an R-homomorphism ϕ: $R[[Z_1,\ldots,Z_s]] \to R[[X_1,\ldots,X_m,Y_1,\ldots,Y_n]]$ by $\phi(Z_i) = X_i + f_i$ for $1 \leq i \leq s$. Such a morphism does exist by Theorem 44.

As in the proof of Theorem 44, let ψ be the R-homomorphism from $R[[X_1,\ldots,X_m,Y_1,\ldots,Y_n]]$ into $R[[Y_1,\ldots,Y_n]]$ defined by $\psi(\Sigma_{i=0}^{\infty} g_i Y^i) = \Sigma_{i=0}^{\infty} \sigma(g_i) Y^i$. Here we are assuming that $g_i \in R[[X_1, \ldots,X_m]]$. Note that ψ leaves every element of the subring $R[[Y_1, \ldots,Y_n]]$ fixed. The composition, $\psi \circ \phi$, is an R-homomorphism from $R[[Z_1,\ldots,Z_s]]$ into $R[[Y_1,\ldots,Y_n]]$ having the property that

$$(\psi \circ \phi)(Z_i) = \psi(X_i+f_i) = \sigma(X_i) + f_i = a_i + f_i \text{ for } 1 \leq i \leq s$$

The "change of name" isomorphism between $R[[X_1,\ldots,X_s]]$ and $R[[Z_1, \ldots,Z_s]]$ can now be applied to complete the proof.

Theorems 49, 50, and 51 can be nicely summarized.

THEOREM 52. Let R be a ring and let $X_1, \ldots, X_m, Y_1, \ldots, Y_n$ be indeterminates of R. Let $a_1, \ldots, a_m \in R$ and let $f_1, \ldots, f_m$ be elements of $(Y_1,\ldots,Y_n)R[[Y_1,\ldots,Y_n]]$. Then there exists an

R-homomorphism ϕ: $R[[X_1,\ldots,X_m]] \to R[[Y_1,\ldots,Y_n]]$ such that $\phi(X_i) = a_i + f_i$ for $1 \le i \le m$ if and only if $a_i \in I_c(R)$ for all i. Moreover, when such a homomorphism exists, it is unique.

REMARK. The net effect of Theorem 52 is the following. Suppose that S is a ring and that $X_1, \ldots, X_m, Y_1, \ldots, Y_n$ are indeterminates. Let $c_1 + f_1, \ldots, c_m + f_m$ be elements of $S[[Y_1,\ldots, Y_n]]$ where $c_i \in S$ and $f_i \in (Y_1,\ldots,Y_n)$. The existence of a necessarily unique S-homomorphism between $S[[X_1,\ldots,X_m]]$ and $S[[Y_1,\ldots, Y_n]]$ mapping X_i to $c_i + f_i$ for every i depends only on whether or not $c_i \in I_c(S)$ for all i.

Consequently, for a given ring S it becomes important, for the problems under discussion, to determine the ideal $I_c(S)$. Now Theorem 50 says that to compute $I_c(S)$, we have only to consider the case of a single variable and this case was substantially treated by M. O'Malley in [25]. O'Malley's results must await the next section for we conclude this section with two more applications of the Eakin-Sathaye Theorem.

It is a well-known fact that any surjective R-endomorphism of $R[X_1,\ldots,X_m]$ is an automorphism. This result is also true for any power series ring $R[[X_1,\ldots,X_m]]$. It has been proven independently by Gilmer and O'Malley [21] and by Eakin and Sathaye [12]. Our proof includes ideas from both.

THEOREM 53. Let R be a ring with $X_1, \ldots, X_m$ indeterminates of R. An R-endomorphism σ: $R[[X_1,\ldots,X_m]] \to R[[X_1,\ldots,X_m]]$ is surjective if and only if it is an automorphism.

Proof. Suppose that σ is surjective and that $\sigma(X_i) = c_i + f_i$, where $c_i \in R$, $f_i \in (X_1,\ldots,X_m)$. Then by Theorem 48 there exists an R-automorphism τ of $R[[X_1,\ldots,X_m]]$ such that $\tau(X_i) = c_i + X_i$. Thus, $(\sigma \circ \tau^{-1})(X_i) = f_i$. Consequently, since τ^{-1} is an automorphism, if $\sigma \circ \tau^{-1}$ is injective, so is σ. Hence, we may assume that $\sigma(X_i) \in (X_1,\ldots,X_m)$ for each i.

Therefore, we may write

$$\sigma(X_i) = \ell_i + g_i$$

where ℓ_i is a linear form in $X_1, \ldots, X_m$ and $g_i \in (X_i,\ldots,X_m)^2$. Now σ is surjective and so for $1 \leq i \leq m$ there exists a power series h_i such that $\sigma(h_i) = X_i$. Since σ is R-linear, the order of h_i is greater than or equal to 1. Write

$$h_i = a_{i1}X_1 + \cdots + a_{im}X_m + k_i$$

where $k_i \in (X_1,\ldots,X_m)^2$. Then

$$\begin{aligned} X_i &= \sigma(h_i) = a_{i1}\sigma(X_1) + \cdots + a_{im}\sigma(X_m) + \sigma(k_i) \\ &= a_{i1}(\ell_1+g_1) + \cdots + a_{im}(\ell_m+g_m) + \sigma(k_i) \\ &= a_{i1}\ell_1 + \cdots + a_{im}\ell_m + \text{(higher order terms)} \end{aligned}$$

It follows that $\ell_1, \ldots, \ell_m$ are linear forms in $X_1, \ldots, X_m$ which generate the free R-module $RX_1 + \cdots + RX_m$.

In fact, if $aX_1^{e_1}\cdots X_m^{e_m}$ is a monomial of degree s, then $aX_1^{e_1}\cdots X_m^{e_m} = a(\Sigma_{j=1}^m a_{1j}\ell_j)^{e_1}\cdots(\Sigma_{j=1}^m a_{mj}\ell_j)^{e_m}$, which is a form in $\ell_1, \ldots, \ell_m$ of degree s. Therefore, the free R-module of all monomials of degree s in the variables $X_1, \ldots, X_m$ is generated by the set $L = \{\ell_1^{e_1}\cdots\ell_m^{e_m} \mid e_1 + \cdots + e_m = s\}$. Since this set has the same cardinality as the basis $\{X_1^{e_1}\cdots X_m^{e_m} \mid e_1 + \cdots + e_m = s\}$, it follows from linear algebra that the set L is also free. In particular, if h_s is a nonzero form of degree s in the variables $X_1, \ldots, X_m$, then $h_s(\ell_1,\ldots,\ell_m)$ is different from zero.

We now prove that σ is injective. Let h be a nonzero power series of order s, $h = h_s + h'$, where h_s is a form of degree s and $h' \in (X_1,\ldots,X_m)^{s+1}$. Then

$$\sigma(h) = \sigma(h_s) + \sigma(h') = h_s(\ell_1+g_1,\ldots,\ell_m+g_m) + \sigma(h')$$

where $\sigma(h')$ has order greater than s. From this we see that $h_s(\ell_1,\ldots,\ell_m)$ is the initial term of $\sigma(h)$ and, as proved above, $h_s(\ell_1,\ldots,\ell_m) \neq 0$. Therefore, $\sigma(h) \neq 0$ and σ is injective.

We are also in a position to determine the automorphisms of $R[[X_1,\ldots,X_m]]$ at least in so far as we are able to determine $I_c(R)$. The ideas are already present in our proof of Theorem 53 and the result itself may be found in [21].

THEOREM 54. Let R be a ring with X_1, ..., X_m indeterminates of R. Let c_1, ..., c_m be elements of $I_c(R)$ and let f_1, ..., f_m be elements of $(X_1,\ldots,X_m)$ where $f_i = \ell_i + g_i$, with ℓ_i a linear form in X_1, ..., X_m and g_i an element of $(X_1,\ldots,X_m)^2$. Then the R-endomorphism σ of $R[[X_1,\ldots,X_m]]$ determined by the conditions $\sigma(X_i) = c_i + f_i$ for all i is an automorphism if and only if the determinant of the matrix of coefficients of the ℓ_i's is a unit in R.

Proof. We begin exactly the way we began the proof of Theorem 53. Thus, we may assume that $c_i = 0$ for $1 \leq i \leq m$.

If σ is an automorphism, then σ is surjective and so the linear forms ℓ_1, ..., ℓ_m generate the free R-module $RX_1 + \cdots + RX_m$. It is a well-known fact from linear algebra that this holds if and only if the matrix of coefficients of the ℓ_i's is invertible, which holds if and only if the determinant of the matrix of coefficients of the ℓ_i's is a unit of R.

Conversely, if the determinant of the matrix of coefficients of the ℓ_i's is a unit, then the linear forms ℓ_1, ..., ℓ_m generate the free R-module $RX_1 + \cdots + RX_m$. Therefore, as argued in the proof of Theorem 53, the R-module of all monomials of degree s in the variables X_1, ..., X_m is generated by the set $\{\ell_1^{e_1}\cdots\ell_m^{e_m} \mid e_1 + \cdots + e_m = s\}$ for any positive integer s. Consequently, if H_s is a form of degree s, there exists a form, G_s, of degree s such that $G_s(\ell_1,\ldots,\ell_m) = H_s$.

Let $F = F_0 + F_1 + \cdots$ be any power series in $R[[X_1,\ldots,X_m]]$ where F_i is a form of degree i. Now we are assuming that $c_i = 0$ for $1 \le i \le m$ and so σ is "substitution by $f_1, \ldots, f_m$." We prove by induction that there exists a sequence $\{G_i\}_{i=0}^{\infty}$ of forms of degree i such that for any positive integer n, $(F - \Sigma_{i=0}^{n} G_i(f_1,\ldots, f_m)) \in (X_1,\ldots,X_m)^{n+1}$. Set $G_0 = F_0$ and having appropriately defined $G_0, \ldots, G_n$, we seek to find G_{n+1}. Let H_s be the leading form of $F - \Sigma_{i=0}^{n} G_i(f_1,\ldots,f_m)$. If $s > n + 1$, let $G_{n+1} = 0$. If $s = n + 1$, as argued above, we can find a form, G_{n+1}, of degree $n + 1$ such that $G_{n+1}(\ell_1,\ldots,\ell_m) = H_{n+1}$. In any event, $F - \Sigma_{i=0}^{n+1} G_i(f_1, \ldots,f_m)$ belongs to $(X_1,\ldots,X_m)^{n+2}$ and the induction is complete.

Thus, set $G = \lim_{n\to\infty}(G_0+\cdots+G_n) = \Sigma_{i=0}^{\infty} G_i$. Then σ is continuous and so $\sigma(G) = \lim_{n\to\infty}\sigma(G_0+\cdots+G_n) = G$, the continuity and limit being taken with respect to the $(X_1,\ldots,X_m)$-adic topology on $R[[X_1,\ldots,X_m]]$. It follows that σ is surjective. By Theorem 53, σ being surjective is equivalent to σ being an automorphism.

13. Computing $I_c(R)$

Let R be a ring. We saw in the preceding section that computation of the ideal $I_c(R)$ was the essential factor in determining which power series could be the images of the variables under homomorphisms. In this section we shall indicate how to decide whether a particular element a of R does or does not belong to $I_c(R)$, at least in case $\bigcap_{n=1}^{\infty} a^n R = (0)$. We can say a little in case $\bigcap_{n=1}^{\infty} a^n R \neq (0)$ and that we shall do at the end of this section. For the time being, assume that $\bigcap_{n=1}^{\infty} a^n R = (0)$. By Theorem 50, we must decide whether or not there exists an R-homomorphism ϕ: $R[[X]] \to R$ such that $\phi(X) = a$. If such a morphism does exist, we must have that $\phi(r_0+r_1X+\cdots+r_mX^m) = r_0 + r_1a + \cdots + r_ma^m$. Moreover, we note that if $\Sigma_{i=0}^{\infty} r_iX^i$ is any power series, then the sequence $\{\Sigma_{i=0}^{m} r_ia^i\}_{m=0}^{\infty}$ is a Cauchy sequence of R in the aR-adic topology. Since $\bigcap_{n=1}^{\infty} a^n R = (0)$, R is a Hausdorff space and so, if the above sequence has a limit, it has a unique limit. Thus, if R is complete in the aR-adic topology, all Cauchy sequences have limits and we could define

$$\phi\left(\sum_{i=0}^{\infty} r_i X^i\right) = \lim_{m\to\infty}\left\{\sum_{i=0}^{m} r_i a^i\right\}_{m=0}^{\infty}$$

In this section we shall show in Theorem 54 that in case R is complete in the aR-adic topology, this definition determines an R-homomorphism ϕ from R[[X]] onto R such that $\phi(X) = a$. In other words, $a \in I_c(R)$. Surprisingly, the converse is also true. That is, if $a \in I_c(R)$ with $\cap_{n=1}^{\infty} a^n R = (0)$ then R is complete in its aR-adic topology. This we shall also prove. We begin with the first of these results.

THEOREM 55. Let R be a ring with a an element of R. Suppose that $\cap_{n=1}^{\infty} a^n R = (0)$ and that R is complete in the aR-adic topology. For $f = \Sigma_{i=0}^{\infty} r_i X^i \in R[[X]]$, define

$$\phi(f) = \lim_{m\to\infty}\left\{\sum_{i=0}^{m} r_i a^i\right\}_{m=0}^{\infty} \underline{\underline{\text{def.}}} \sum_{i=0}^{\infty} r_i a^i$$

Then ϕ is an R-homomorphism from R[[X]] onto R and $\phi(X) = a$.

Proof. As noted previously, $\{\Sigma_{i=0}^{m} r_i a^i\}_{m=0}^{\infty}$ is a Cauchy sequence for any element $f = \Sigma_{i=0}^{\infty} r_i X^i$ of R[[X]]. Thus, $\phi(f) = \Sigma_{i=0}^{\infty} r_i a^i$ is well defined. Let $f = \Sigma_{i=0}^{\infty} r_i X^i$ and $g = \Sigma_{i=0}^{\infty} s_i X^i$ be elements of R[[X]] with r,s elements of R. By definition, $\phi(rf + sg) = \Sigma_{i=0}^{\infty}(rr_i + ss_i)a^i$ and $r\phi(f) + s\phi(g) = r\,\Sigma_{i=0}^{\infty} r_i a^i + s\,\Sigma_{i=0}^{\infty} s_i a^i$. To prove their equality we have, since R is a topological ring, to show that the sequence

$$\left\{\left[\left(r \sum_{i=0}^{m} r_i a^i\right) + \left(s \sum_{i=0}^{m} s_i a^i\right)\right] - \left[\sum_{i=0}^{m} (rr_i + ss_i)a^i\right]\right\}_{m=0}^{\infty}$$

converges to zero. This, however, is obvious and therefore ϕ is R-linear.

We now prove that ϕ is a ring morphism. Again we shall make use of the fact that R is a topological ring with respect to its aR-adic topology. We must prove that if $f = \Sigma_{i=0}^{\infty} r_i X^i$ and $g = \Sigma_{i=0}^{\infty} s_i X^i$, then $(\Sigma_{i=0}^{\infty} r_i a^i)(\Sigma_{i=0}^{\infty} s_i a^i) = \Sigma_{i=0}^{\infty} t_i a^i$, where $t_i = \Sigma_{j+k=i} r_j s_k$. Now

$$\left(\sum_{i=0}^{\infty} r_i a^i\right)\left(\sum_{i=0}^{\infty} s_i a^i\right) = \left(\lim_{m\to\infty} \sum_{i=0}^{m} r_i a^i\right)\left(\lim_{m\to\infty} \sum_{i=0}^{m} s_i a^i\right)$$

$$= \lim_{m\to\infty}\left[\left(\sum_{i=0}^{m} r_i a^i\right)\left(\sum_{i=0}^{m} s_i a^i\right)\right]$$

Thus, we have only to see that

$$\left\{\left(\sum_{i=0}^{m} r_i a^i\right)\left(\sum_{i=0}^{m} s_i a^i\right) - \left(\sum_{i=0}^{m} t_i a^i\right)\right\}_{m=0}^{\infty}$$

is a null sequence, where $t_i = \Sigma_{j+k=i} r_j s_k$. But for any m,

$$\left(\sum_{i=0}^{m} r_i a^i\right)\left(\sum_{i=0}^{m} s_i a^i\right) = \sum_{i=0}^{m} t_i a^i + (\text{terms of degree in } a \geq m)$$

from which the assertion follows.

We turn at once to the converse.

THEOREM 56. Let R be a ring with a an element of R. Suppose that $\cap_{n=1}^{\infty} a^n R = (0)$ and that there exists an R-homomorphism ϕ: $R[[X]] \to R$ such that $\phi(X) = a$. Then R is complete in its aR-adic topology.

Proof. Let $\{c_m\}_{m=0}^{\infty}$ be a sequence of elements of R which is Cauchy in the aR-adic topology. It is routine to show that there exists a subsequence of $\{c_m\}_{m=0}^{\infty}$ of the form $\{\Sigma_{i=0}^{m} r_i a^i\}_{m=0}^{\infty}$ where $r_i \in R$. [See Exercise 5.]

Since the aR-adic topology is Hausdorff,, if limits exist, they are unique and we have only to show that $\{\Sigma_{i=0}^{m} r_i a^i\}_{m=0}^{\infty}$ has a limit in R. It is clear that ϕ is a continuous homomorphism from R[[X]] in the (X)-adic topology to R in the aR-adic topology for we have only to check continuity at zero. This is easy for if $a^n R$ is a basic neighborhood of zero in R, then (X^n) is a neighborhood of zero in R[[X]] and $\phi(X^n) \subseteq a^n R$.

Now if $f = \Sigma_{i=0}^{\infty} r_i X^i$, then $f \in R[[X]]$ and so $\phi(f) \in R$. Moreover, $f = \lim_{m\to\infty}(\Sigma_{i=0}^{m} r_i X^i)$ and so, by the continuity of ϕ,

$$\phi(f) = \lim_{m\to\infty} \phi\left(\sum_{i=0}^{m} r_i X^i\right) = \lim_{m\to\infty} \left(\sum_{i=0}^{m} r_i a^i\right)$$

Combining Theorems 55 and 56 we obtain the following result, the first proof of which was given by O'Malley in [25].

THEOREM 57. Let R be a ring with a an element of R and suppose that $\bigcap_{n=1}^{\infty} a^n R = (0)$. Then $a \in I_c(R)$ if and only if R is complete in the aR-adic topology on R.

Thus, in the Hausdorff case we have a very precise test for determining whether or not an element belongs to I_c. No such test has yet been found in the nonHausdorff situation and one might even hope that the situation would never arise. Not so however, for we next present an example, due to Gilmer [20], of a ring R containing an element a such that $\bigcap_{n=1}^{\infty} a^n R \neq (0)$ but $a \in I_c(R)$.

Example. This is an example of a ring R containing an element a such that $a \in I_c(R)$, but $\bigcap_{n=1}^{\infty} a^n R \neq (0)$.

Let R be a ring with X an indeterminate. Suppose that b is an element of R and ϕ_b is an R-homomorphism from R[[X]] onto R with $\phi_b(X) = b$. If I is an ideal of R such that $\phi_b(I[[X]]) \subseteq I$, then

ϕ_b induces an R-homomorphism from $R[[X]]/I[[X]]$ onto R/I. Since $(R[[X]]/I[[X]]) \simeq (R/I)[[X]]$, this in turn induces an (R/I)-homomorphism on $(R/I)[[X]]$ mapping X to $b + I$. Moreover, the element $b + I$ determines a Hausdorff topology on R/I if and only if $\cap_{n=1}^{\infty}(b^n + I) = I$ and it is a known fact that $\cap_{n=1}^{\infty}(b^n + I)$ is the closure of I in the bR-adic topology on R [34].

It follows from these observations that if we start with a ring S and an element b in $I_c(S)$, either b is our example or we can choose an ideal I of S such that $\phi_b(I[[X]]) \subseteq I$ and try S/I. Now the condition $\phi_b(I[[X]] \subseteq I$ need not always hold, but if I is a principal ideal, say αS, then $I[[X]] = (\alpha S)[[X]]$ and if $\alpha f \in (\alpha S)[[X]]$, then

$$\phi_b(\alpha f) = \alpha\phi_b(f) \in \alpha S$$

Therefore, the condition is always satisfied by principal ideals of S.

Let S be a ring with b an element of S and Y an indeterminate. Set $A = \cap_{n=1}^{\infty} b^n S$ and consider the power series $b + Y$. We claim that if $A \subseteq (b + Y)S[[Y]]$, then $bA = A$. Clearly, we have only to prove that $A \subseteq bA$. If $t \in A$, then $t = (b + Y)(\Sigma_{i=0}^{\infty} s_i Y^i)$ for some $s_i \in S$. Therefore, the following system of equations is valid.

$$\begin{aligned} t &= bs_0 \\ 0 &= s_0 + bs_1 \\ &\vdots \\ 0 &= s_m + bs_{m+1} \end{aligned}$$

It is obvious from these equations that $s_m \in bS$ for $m \geq 0$. Suppose that $s_i \in b^n S$ for $i \geq 0$. Then $s_m = b(-s_{m+1}) \in b^{n+1} S$ for $m \geq 0$. It follows by induction that $s_i \in \cap_{n=1}^{\infty} b^n S = A$ for $i \geq 0$ and, in particular $s_0 \in A$. Hence, $t = bs_0 \in bA$.

Now suppose that S is a ring containing an element b such that $b[\cap_{n=1}^{\infty} b^n S] \subseteq \cap_{n=1}^{\infty} b^n S$. Let Y be an indeterminate and consider the power series b + Y. Then we claim that $(b + Y)S[[Y]]$ is not closed in the (Y)-adic topology on S[[Y]]. By the previous observation, $\cap_{n=1}^{\infty} b^n S \subsetneq (b + Y)S[[Y]]$ and, as noted earlier, it is a fact [34] that the closure of $(b + Y)S[[Y]]$ is $\cap_{n=1}^{\infty}(b + Y, Y^n)$. Therefore, we will have justified the claim as soon as we can show that $\cap_{n=1}^{\infty} b^n S \subseteq \cap_{n=1}^{\infty}(b + Y, Y^n)$.

Since either intersection may be taken over even indices only, it is obviously sufficient to show that for each even integer m, $b^m \in (b + Y, Y^m)$. But

$$b^m = b^m - Y^m + Y^m \in (b + Y, Y^m)$$

So, let S be a ring containing an element b such that $b[\cap_{n=1}^{\infty} b^n S] \subsetneq \cap_{n=1}^{\infty} b^n S$ and let X and Y be indeterminates. There exists an (S[[Y]])-homomorphism ϕ_Y: $(S[[Y]])[[X]] \to S[[Y]]$ with $\phi_Y(X) = Y$. As we have just shown, the element b + Y of S[[Y]] is such that the ideal $I = (b + Y)S[[Y]]$ is not closed in the (Y)-adic topology of S[[Y]]. Therefore, if we set $R = (S[[Y]])/I$, then by our earlier remarks, ϕ_Y induces an R-homomorphism, written $\bar{\phi}_Y$, from R[[X]] onto R such that $\bar{\phi}_Y(X) = Y + I$. Moreover, $\cap_{n=1}^{\infty}(Y^n + I)$ is nonzero. Summarizing, $Y + I \in I_c(R)$, but $\cap_{n=1}^{\infty}(Y + I)^n$ is different from zero.

Consequently, we have reduced our task to that of finding a ring S containing an element b with the property that $b[\cap_{n=1}^{\infty} b^n S] \subsetneq \cap_{n=1}^{\infty} b^n S$. We shall indicate how to do this while leaving the details as an exercise. Let T be a ring with Y and $\{X_i\}_{i=0}^{\infty}$ indeterminates. Set $S = T[Y, \{X_i\}_{i=0}^{\infty}]/B$ where B is the ideal of $T[Y, \{X_i\}_{i=1}^{\infty}]$ generated by the set $\{X_0 - X_i Y^i\}_{i=1}^{\infty}$. Then the element $y = Y + B$ is such that $x_0 = X_0 + B \in \cap_{n=1}^{\infty} y^n S$, while $x_0 \notin y[\cap_{n=1}^{\infty} y^n S]$. By passing to preimages under the natural map from $T[Y, \{X_i\}_{i=0}^{\infty}]$ onto S, one has only to verify the following three relations.

(1) $\cap_{n=1}^{\infty}[B + (Y^n)] = (X_0, X_1 Y, X_2 Y^2, \ldots)$;

(2) $B + Y\{\cap_{n=1}^{\infty}[B + (Y^n)]\} = (X_0 Y, X_0 - X_1 Y, X_0 - X_2 Y^2, \ldots)$;

(3) $X_0 \notin (X_0Y, X_0-X_1Y, \ldots, X_0-X_nY^n, \ldots)$.

We conclude this section with an interesting theorem of O'Malley [25]. It shows that in a sense, rings of the type constructed above are the only rings which are pathological in the present context.

THEOREM 58. Let R be a ring and suppose that for each element a in the Jacobson radical, $a[\cap_{n=1}^{\infty} a^nR] = \cap_{n=1}^{\infty} a^nR$. Then an element b of R belongs to $I_c(R)$ if and only if R is a complete Hausdorff space in its bR-adic topology.

Proof. If R is a complete Hausdorff space in its bR-adic topology, then $b \in I_c(R)$ by Theorem 55.

Conversely, suppose that $b \in I_c(R)$. Then by Theorems 50 and 47 the R-endomorphism ϕ of R[[X]] mapping X to b + X is an automorphism. Thus, by invoking Exercise 8, we have that

$$\bigcap_{n=1}^{\infty} (b + X)^n = \bigcap_{n=1}^{\infty} (\phi(X))^n = \phi\left(\bigcap_{n=1}^{\infty} (X^n)\right) = \phi((0)) = (0)$$

Now, we have only to show that $\cap_{n=1}^{\infty} b^nR = (0)$, for then the result will follow from Theorems 50 and 56. Hence, in light of the above, it is sufficient to show that $\cap_{n=1}^{\infty} b^nR \subseteq \cap_{n=1}^{\infty} (b + X)^n$.

Therefore, set $\cap_{n=1}^{\infty} b^nR = A$. By hypothesis, bA = A and we wish to see that $A \subseteq \cap_{n=1}^{\infty} (b + X)^n$. We prove the following more general statement. Suppose that $\beta = \Sigma_{i=0}^{\infty} b_iX^i$ is a power series having constant term b. Then bA = A implies that $A \subseteq \cap_{n=1}^{\infty} (\beta^n)$.

We first note that it suffices to prove that if bA = A, then $A \subseteq (\beta)$. The reason is that since bA = A, $b^mA = A$ for each positive integer m. Therefore, $A \subseteq (\gamma)$ for any power series γ having constant term b^m and in particular, $A \subseteq (\beta^m)$.

Thus, we have to show that because bA = A, $A \subseteq (\beta)$. Let $t \in A$. We wish to find a power series

$$\alpha = \sum_{i=0}^{\infty} c_i X^i \qquad \text{such that } \alpha\beta = t$$

Therefore, we wish to find elements c_i of R such that the following system of equations holds simultaneously.

$$c_0 b = t$$
$$c_0 b_1 + c_1 b = 0$$
$$\vdots$$
$$c_0 b_m + \cdots + c_m b = 0$$
$$\vdots$$

Now $t \in A = bA$ and hence there exists an element r of A such that $t = br$. Set $c_0 = r$. Since $c_0 \in A$, there exists an element d_0 of A such that $c_0 = bd_0$. Therefore, if we set $c_1 = -d_0 b_1$, then it follows that $c_0 b_1 + c_1 b = 0$ and further that $c_1 \in A$.

Suppose that for $0 \le i \le m$ we have determined c_i such that $c_i \in A$ and the first $m + 1$ of the above equations are satisfied. For $0 \le i \le m$, write $c_i = bs_i$, $s_i \in A$. If we set $c_{m+1} = -(s_0 b_{m+1} + \cdots + s_m b_1)$, then clearly $c_{m+1} \in A$ and the equation

$$\begin{aligned} c_0 b_{m+1} + c_1 b_m + \cdots + c_{m+1} b &= (c_0 b_{m+1} + c_1 b_m + \cdots + c_m b_1) \\ &\quad - (s_0 b_{m+1} + \cdots + s_m b_1 b) \\ &= (bs_0 b_{m+1} + bs_1 b_m + \cdots + bs_m b_1) \\ &\quad - (s_0 b_{m+1} b + \cdots + s_m b_1 b) \\ &= 0 \end{aligned}$$

says that the $(m + 2)$nd equation of the above is satisfied. By induction, the power series α exists and the result follows.

It is not difficult to show (see Exercise 9) that if R is an integral domain or a Noetherian ring, then R has the property that for each element a of R, $a[\cap_{n=1}^{\infty} a^n R] = \cap_{n=1}^{\infty} a^n R$. This yields the following corollary to Theorem 58.

THEOREM 59. Suppose that R is either an integral domain or a Noetherian ring. Then $I_c(R) = \{a \in R \mid R$ is a complete Hausdorff space in its aR-adic topology$\}$.

It seems to us as if we have inadvertently done an injustice to the work of O'Malley. As presented in this chapter, his papers have largely been ignored and his results improved or proved by other means. This was for the sake of clarity and efficiency and was not for the purpose of denying him his due. It is no great feat to give a more polished treatment of the work of others after time has prepared the way. In so far as this chapter is interesting mathematics, and we find it very interesting, O'Malley deserves the lion's share of the credit for its development.

Exercises

1. Let R be a ring with $X_1, \ldots, X_m, Z_1, \ldots, Z_n$ indeterminates. Suppose that $\sigma: R[[X_1,\ldots,X_m]] \to R$ is an R-homomorphism. Regard $R[[X_1,\ldots,X_m,Z_1,\ldots,Z_n]]$ as $(R[[X_1,\ldots,X_m]])[[Z_1,\ldots, Z_n]]$ and, for notational purpose, write elements of $(R[[X_1,\ldots, X_m]])[[Z_1,\ldots,Z_n]]$ as $\Sigma_{i=0}^{\infty} g_i Z^i$, where $g_i \in R[[X_1,\ldots,X_m]]$ and Z^i denotes a homogeneous polynomial of degree i in $Z_1, \ldots, Z_n$. Prove that the map ϕ determined by $\phi(\Sigma_{i=0}^{\infty} g_i Z^i) = \Sigma_{i=0}^{\infty} \sigma(g_i) Z^i$ is an R-homomorphism from $(R[[X_1,\ldots,X_m]])[[Z_1,\ldots,Z_n]]$ into $R[[Z_1,\ldots,Z_n]]$.

2. Let R be a ring with $X_1, \ldots, X_m$ indeterminates. Prove that there exists an R-endomorphism δ of $R[[X_1,\ldots,X_m]]$ such that $\delta(X_i) = -X_i$ for $1 \le i \le m$. In fact, show that δ is an R-automorphism.

3. If R is a ring, recall that the Jacobson radical of R is the intersection of all maximal ideals of R. Prove that if J is the Jacobson radical of R, then $J + (X_1,\ldots,X_m)$ is the Jacobson radical of $R[[X_1,\ldots,X_m]]$. *Hint:* See Chapter I, Exercise 1.

4. Let R be a ring with a a nilpotent element of R. Prove that there exists an R-homomorphism ϕ of $R[[X]]$ onto R such that $\phi(X) = a$. Conclude that the nilradical of R is contained in $I_c(R)$.

5. Let R be a ring with a an element of R. Suppose that $\cap_{n=1}^{\infty} a^n R = (0)$ and let $\{c_m\}_{m=0}^{\infty}$ be a Cauchy sequence of R in the aR-adic topology. Prove that $\{c_m\}_{m=0}^{\infty}$ contains a subsequence of the form $\{\Sigma_{i=0}^{m} r_i a^i\}_{m=0}^{\infty}$ for some $r_i \in R$.

6. Justify the claim made in the introduction to this chapter. Precisely, if Z is the ring of integers, prove that there does not exist a Z-homomorphism ϕ of $Z[[X]]$ such that $\phi(X) = 1 + X$. Show, in fact, that this holds for any ring R. Also, what is $I_c(Z)$?

7. Prove that the relations (1), (2), and (3) of the example are valid.

8. Let ϕ be an R-automorphism of $R[[X]]$. Prove that

$$\bigcap_{n=1}^{\infty} (\phi(X))^n = \phi\left(\bigcap_{n=1}^{\infty} (X)^n\right) = \phi((0)) = (0)$$

9. Suppose that R is either an integral domain or a Noetherian ring. Prove that for each element a of R, $a[\cap_{n=1}^{\infty} a^n R] = \cap_{n=1}^{\infty} a^n R$.

10. In this exercise we shall demonstrate the fact that $I_c(R)$ may properly contain the nilradical of R as well as be properly contained in the Jacobson radical of R. Prove that this is indeed the case in the following situation: Z, the ring of integers, X and Y indeterminates, $R' = (Z/4Z)[X]$, $M = (2,X)R'$, and $R = R'_M[[Y]]$.

REFERENCES

1. J. Arnold, On Krull dimension in power series rings, *Trans. Amer. Math. Soc.* 177 (1973), 299-304.
2. J. Arnold, Power series rings over Prüfer domains, *Pac. J. Math.* 44 (1973), 1-11.
3. J. Arnold, Prime ideals in power series rings, *Proc. Kansas Commutative Algebra Conference*, Springer-Verlag Lecture Notes in Math. No. 311, Berlin, 1973, 17-25.
4. J. Arnold and J. Brewer, On when $(D[[X]])_{P[[X]]}$ is a valuation ring, *Proc. Amer. Math. Soc.* 37 (1973), 326-332.
5. J. Arnold, R. Gilmer, and W. Heinzer, Some countability conditions in a commutative ring, *Ill. J. Math.* 21 (1977), 648-665.
6. M. Atiyah and I. MacDonald, *Introduction to Commutative Algebra*, Addison-Wesley, 1969.
7. N. Bourbaki, *Commutative Algebra*, Addison-Wesley, Reading, Mass., 1972.
8. J. Brewer, E. Rutter, and J. Watkins, Coherence and weak global dimension of R[[X]] when R is von Neumann regular, *J. Alg.* 46 (1977), 278-289.
9. J. B. Castillon, Groupe fini d'automorphismes des anneaux de polynômes et de series formelles, *Bull. Sci. Math.* (2) 95 (1971), 237-240.

10. J. B. Castillon, Anneaux d'invariants de groupes finis d'automorphismes d'algebres de series formelles, *C. R. Acad. Sci. Paris Sci. A* 275 (1972), 627-628.

11. J. B. Castillon, Reciproque a un theoreme de P. Samuel, *C. R. Acad. Sci. Paris Sci. A* 283 (1976), 141-142.

12. P. Eakin and A. Sathaye, R-endomorphisms of R[[X]] are essentially continuous, *Pac. J. Math.* 66 (1976), 83-87.

13. D. Fields, Zero divisors and nilpotent elements in power series rings, *Proc. Amer. Math. Soc.* 27 (1971), 427-433.

14. D. Fields, Dimension theory in power series rings, *Pac. J. Math.* 35 (1970), 601-611.

15. R. Fossum, *The Divisor Class Group of a Krull Domain*, Springer-Verlag, New York and Berlin, 1973.

16. R. Gilmer, *Multiplicative Ideal Theory*, Marcel Dekker, New York, 1972.

17. R. Gilmer, On polynomial and power series rings over a commutative ring, *Rocky Mt. J. Math.* 5 (1975), 157-175.

18. R. Gilmer, A note on the quotient field of the domain D[[X]], *Proc. Amer. Math. Soc.* 18 (1967), 1138-1140.

19. R. Gilmer, Integral dependence in power series rings, *J. Alg.* 11 (1969), 448-502.

20. R. Gilmer, R-automorphisms of R[[X]], *Mich. Math. J.* 17 (1970), 15-21.

21. R. Gilmer and M. O'Malley, R-endomorphisms of $R[[X_1,\ldots,X_n]]$, *Advances in Math.* (in press).

22. M. Hochster, Non-uniqueness of the ring of coefficients in a polynomial ring, *Proc. Amer. Math. Soc.* 34 (1972), 81-82.

23. I. Kaplansky, *Commutative Rings*, Allyn-Bacon, Boston, 1970.

24. J. Ohm and R. Pendleton, Rings with Noetherian spectrum, *Duke Math. J.* 35 (1968), 631-639.

25. M. O'Malley, R-automorphisms of R[[X]], *Proc. London Math. Soc.* (3) 20 (1970), 69-78.

26. M. O'Malley, Some remarks on the formal power series ring, *Bull. Soc. Math. Fr.* 99 (1971), 247-258.

27. M. O'Malley, Finite groups of R-automorphisms of R[[X]], *Mich. Math. J.* 20 (1973), 277-284.

28. M. O'Malley, Isomorphic power series rings, *Pac. J. Math.* 41 (1972), 503-512.

29. R. Rivet, Sur le corps des fractions d'un anneau de series formelles a coefficients dans un anneau de valuation discrete, *C. R. Acad. Sci. Paris Ser. A* 264 (1967), 1047-1049.

30. P. Samuel, On unique factorization domains, *Ill. J. Math.* 5 (1961), 1-17.

31. P. Samuel, Groupes finis d'automorphismes des anneaux de series formelles, *Bull. Sci. Math. Fr.* (2) 90 (1966), 97-101.

32. P. Sheldon, How changing D[[X]] changes its quotient field, *Trans. Amer. Math. Soc.* 159 (1971), 223-244.

33. O. Zariski and P. Samuel, *Commutative Algebra*, Vol. 1, Van Nostrand, Princeton, N.J., 1958.

34. O. Zariski and P. Samuel, *Commutative Algebra*, Vol. 2, Van Nostrand, Princeton, N.J., 1960.

INDEX